# NANOCRYSTALS – SYNTHESIS, CHARACTERIZATION AND APPLICATIONS

Edited by **Sudheer Neralla**

**Nanocrystals – Synthesis, Characterization and Applications**
http://dx.doi.org/10.5772/2560
Edited by Sudheer Neralla

### Contributors

Noelio Oliveira Dantas, Ernesto Soares de Freitas Neto, Peter Petrik, P. Vengadesh, Ricardo Souza da Silva, Ernesto Soares de Freitas Neto, Noelio Oliveira Dantas, Igor Yu. Denisyuk, Julia A. Burunkova, Sandor Kokenyesi, Vera G. Bulgakova, Mari Iv. Fokina, Chengjun Zhou, Qinglin Wu, Anurag Srivastava, Neha Tyagi, Liang-Yih Chen, Hung-Lung Chou, Ching-Hsiang Chen, Chia-Hung Tseng, Xuejun Zhang, Fuxing Gan

### Published by InTech

Janeza Trdine 9, 51000 Rijeka, Croatia



**Publishing Process Manager** Dragana Manestar
**Technical Editor** InTech DTP team
**Cover** InTech Design team

Additional hard copies can be obtained from orders@intechopen.com

Nanocrystals – Synthesis, Characterization and Applications, Edited by Sudheer Neralla
p. cm.
ISBN 978-953-51-0714-9

For EU product safety concerns:
IN TECH d.o.o., Prolaz Marije Krucifikse Kozulić 3, 51000 Rijeka, Croatia,
info@intechopen.com or visit our website at intechopen.com.

# Meet the editor

Dr Sudheer Neralla is a Research Associate at Jet-Hot LLC, Burlington, NC, USA He received his PhD from North Carolina A&T State University, Greensboro, NC, USA in 2006. His current research work in collaboration with NSF-Engineering Research Center, NC A&T State University, involves mechanical property evaluation of Mg-based biodegradable alloys for craniofacial and orthopedic applications. His doctoral research work was on synthesis of nanoparticles, metal-ceramic nanocomposite thin films using pulsed laser deposition and magnetron sputtering, microstructural and tribological characterization of thin films, nanoindentation of thin films and nanocatalyst synthesis for growth of multi-wall carbon nanotube arrays. His research also includes corrosion resistance and thermal property study of coatings.

# Contents

# Preface

This book provides an overview of the synthesis and characterization of nanocrystals. Nanocrystals, owing to their unique behavior with reduction in size, have been a significant part of the novel materials developed for applications such as biosensors, optics, catalysts to semiconductor devices. Over the years, various synthesis methods are discovered to develop nanostructures with tunable properties such as optical, electronic magnetic and mechanical properties. The chapters in this book cover a broad range of properties of nanocrystals synthesized for various applications.

Chapter 1 discusses optical absorption and photoluminescence properties of $Cd_{1-x}Mn_xS$ nanoparticles grown by the melting-nucleation synthesis approach. The difference in magneto-optical behavior of nanocrystals and quantum dots are discussed. A spectroscopic ellipsometry method used to characterize nanocrystals is described in chapter 2. The basics, measurable nanocrystal properties and range applications of spectroscopic ellipsometry are explained in this chapter. Chapter 3 presents an overview of the density function theory (DFT) software used for the calculations of different periodic and non-periodic systems. Density of crystal structures and spectroscopic properties of nanoparticles are evaluated. In chapter 4, fabrication of photovoltaic device using carboxymethyl cellulose and Chlorophyll-a nanocrystals and Bacteriorhodopsin is explained. Spectroscopic and photoelectric properties are analyzed to evaluate the material suitability. Chapter 5 presents an overview of optical, magnetic and structural properties $Cd_{1-x}Mn_xS$, $Pb_{1-x}Mn_xS$, $Zn_{1-x}Mn_xO$ nanocrystals grown by fusion and co-precipitation methods. Effect of secondary phase on the properties of the nanocrystals is studied using x-ray diffraction and Raman spectroscopic analysis of the nanocrystals. Chapter 6 describes UV-curable nanocomposite materials with self-writing properties like light self-focusing and light induced nanoparticle redistribution. Mechanical, optical properties of these ZnO based nanocomposites are studied and explained in detail. The developments in the applications of cellulose nanocrystals (CNC) in nanocomposites prepared by gelation and electrospinning are reported in chapter 7. Nanocomposite fibers containing CNC are synthesized using electrospinning. Chapter 8 discusses various applications of semiconductor nanocrystals, their synthesis and electronic, structural, optical, magnetic and mechanical properties. Structural transformation of nanocrystals under pressure is studied. Chapter 9 presents an overview of surface modification of colloidal semiconductor CdS and CdSe quantum dots using organic ligands and their characterization using time-resolved photoluminescence (TRPL) spectroscopy and density function theory (DFT).

I believe our contribution provides a significant value to the science and technology community resulting in more discoveries in diverse fields implementing nanotechnology.

**Dr. Sudheer Neralla**
NSF-Engineering Research Center
North Carolina A&T State University
Greensboro,
USA

Chapter 1

# Carrier Dynamics and Magneto-Optical Properties of $Cd_{1-x}Mn_xS$ Nanoparticles

Noelio Oliveira Dantas and Ernesto Soares de Freitas Neto

Additional information is available at the end of the chapter

http://dx.doi.org/10.5772/46515

## 1. Introduction

$Cd_{1-x}Mn_xS$ nanoparticles (NPs) with size quantum confinement belong to the diluted magnetic semiconductor (DMS) quantum dot (QD) class of materials that has been widely studied in the last few years. The study of quasi-zero-dimensional Diluted Magnetic Semiconductors (DMS), such as $Cd_{1-x}Mn_xS$ Quantum Dots (QDs), is strongly motivated due to the localization of magnetic ions in the same places as the free-like electron and hole carriers occurring in these nanomaterials **[1,2]**. This interesting phenomenon causes unique properties in DMS dots that can be explored in different technological applications, such as wavelength tunable lasers**[3]**, solar cells**[4,5]**, or in spintronic devices**[6,7]**. In this context, glass matrix-encapsulated $Cd_{1-x}Mn_xS$ NPs emerge as potential candidates for several applications, given that this host transparent material is robust and provides excellent stability for DMS nanostructures. Therefore, the luminescent properties and carrier dynamics of $Cd_{1-x}Mn_xS$ NPs should be comprehensively understood in order to target optical applications. For instance, different models based on rate equations can be employed to describe the temperature-dependent carrier dynamics of DMS nanostructures, such as they have been applied to semiconductor quantum wells**[8]**, N-impurity complexes in III–V materials**[9]**, and self-assembled semiconductor quantum dots**[10]**.

It is well known that the optical properties of NPs can be significantly changed by interactions between nanostructures and their host material, due mainly to the formation of surface defects **[11, 12]**. These surface defects are heavily dependent on NP size and become more important with increasing surface–volume ratio. Generally, the comparison between the optical properties of $Cd_{1-x}Mn_xS$ QDs and their corresponding bulk is obtained in different environments. To the best of our knowledge, this study is probably the first that simultaneously investigates both the carrier dynamics and the magneto-optical properties of $Cd_{1-x}Mn_xS$ QDs and their corresponding bulk-like NC when both are embedded in the same host material.

Although the dot doped with impurities (metal and magnetic) are currently being synthesized by colloidal chemistry techniques **[13,14]**, some possible applications require the nanoparticles (NPs) being embedded in robust and transparent host materials. In this context, the melting-nucleation approach appears as an appropriate synthesis technique since it allows the growth of DMS nanocrystals (NCs) embedded in different glass matrices. In addition to the controllable dot size and $Mn^{2+}$ ion fraction incorporated into $Cd_{1-x}Mn_xS$ dots which can be achieved by this synthesis protocol, for example, the host glass matrix provides an excellent stability to the NPs. In particular for the melting-nucleation protocol used in this chapter, it is presented a discussion on the doping of QDs with magnetic impurities reasoned in two main models**[3]**: the 'trapped-dopant' and 'self-purification' mechanisms.

In this chapter, we have employed the optical absorption (OA), magnetic force microscopy (MFM), photoluminescence (PL), and magnetic circularly polarized photoluminescence (MCPL) measurements in order to investigate the properties of $Cd_{1-x}Mn_xS$ NPs that were successfully grown in a glass matrix. The organization of this chapter is shown as follows. In the section 2 (next section), we present the synthesis protocol that was employed in order to grow $Cd_{1-x}Mn_xS$ NPs in a glass matrix. The results obtained from the experimental techniques are presented and discussed in the section 3, highlighting the carrier dynamics and the magneto-optical properties of nanoparticles. We conclude our study in the section 4.

## 2. Synthesis of $Cd_{1-x}Mn_xS$ nanoparticles in a glass matrix

The host glass matrix for NP growth was labeled SNAB since its nominal composition is: $40\mathbf{Si}O_2.30\mathbf{Na}_2CO_3.1\mathbf{Al}_2O_3.29\mathbf{B}_2O_3$ (mol %). $Cd_{1-x}Mn_xS$ NPs were successfully synthesized in this glass matrix by adding 2[CdO + S] (wt % of SNAB), and x[Mn] (wt % of Cd), with x = 0.0, 0.5, 5.0, and 10 %. The synthesis method consists in a two sequential melting-nucleation approach, in which it is possible obtain ensembles of nearly spherical nanoparticles embedded in a glass matrix **[12]**. First, the powder mixture was melted in an alumina crucible at 1200 ºC for 30 minutes. Next, the melted mixture was quickly cooled down to room temperature where diffusion of $Cd^{2+}$, $Mn^{2+}$, and $S^{2-}$ species took place. This diffusion resulted in $Cd_{1-x}Mn_xS$ NP growth in the SNAB glass environment.

In a second stage, a sample with x = 0.100 was subjected to a thermal annealing at 560 ºC for 6 h in order to enhance the diffusion of ions within the host SNAB matrix which promotes the growth of magnetic dots. Room temperature XRD pattern of the undoped CdS NPs (x = 0) embedded in the SNAB glass matrix was recorded with a XRD-6000 Shimadzu diffractometer using monochromatic Cu-$K_{\alpha 1}$ radiation ($\lambda$ = 1.54056 Å). Thus, the wurtzite structure of CdS NPs embedded in the SNAB glass matrix has been confirmed. Evidently, the $Cd_{1-x}Mn_xS$ NPs with diluted magnetic doping have this same wurtzite structure, since it is a common phase for this DMS material.

## 3. Results and discussions

We have employed several experimental techniques in order to investigate the carrier dynamics and the magneto-optical properties of $Cd_{1-x}Mn_xS$ NPs. The room temperature absorption band edge of synthesized $Cd_{1-x}Mn_xS$ NCs was obtained with a double beam UV – VIS – NIR spectrophotometer (Varian, Cary 500) operating between 250 and 800 nm and with a spectral resolution of 1 nm. Photoluminescence (PL) measurements were taken with a 405 nm (~3.06 eV) continuous wave laser focused on a ~200 μm ray spot with an excitation power of 2.5 mW. $Cd_{1-x}Mn_xS$ NP luminescence was collected using a USB4000 spectrometer from Ocean Optics equipped with a Toshiba TCD1304AP 3648-element linear CCD-array detector, in the 10 K to 300 K temperature range, with a 435 nm high-pass filter. The magnetic force microscopy images of the $Cd_{1-x}Mn_xS$ NPs doped with x = 0.100 were recorded at room temperature with a scanning probe microscope (Shimadzu, SPM – 9600).

The magneto-photoluminescence (MPL) measurements were performed using superconductor coils (Oxford Instruments) with fields up to 15 T. The samples were placed into the liquid helium cryostat at 2 K and excited using a 405 nm (± 5 nm) continuous wave laser, from Laserline Laser Technology, focused on ~ 200 μm rays spot with excitation intensity values of 10 mW. The detected MPL was carried out with an ocean optics spectrometer (USB4000) and the polarization was analyzed using a λ/4 waveplate and with linear polarizer fixed parallel to the spectrometer entrance, in order to collect the photons with $\sigma^+$ and $\sigma^-$ circular polarizations, respectively.

## 3.1. Carrier dynamics

The room temperature OA spectra of $Cd_{1-x}Mn_xS$ NPs, with different x-concentrations, are shown in Fig. 1a. The formation of two well defined groups of $Cd_{1-x}Mn_xS$ NPs of different sizes was confirmed by the two bands in the OA spectra. As indicated in Fig. 1a, these two groups of NPs were named: (i) QDs because their quantum confinement properties provoked a change in band energy around ~3 eV; and (ii) bulk-like NCs indicated by the absence of quantum confinement given the fixed band around ~2.58 eV, a value near the energy gap of bulk CdS **[15,16]**. At the bottom of Fig. 1a is the OA spectrum of the SNAB glass matrix where, in contrast, it can be seen that over a broad spectral range there is a complete absence of any band associated with NPs.

Figure 1a shows that the undoped CdS QDs (x = 0.000) exhibit confinement energy ( $E_{conf}$ ) as indicated by the OA band peak at ~3.10 eV. From this value and using a confinement model based on effective mass approximation**[12,15-18]**, the mean QD radius R was estimated by the expression: $E_{conf} = E_g + (\hbar^2\pi^2 / 2\mu R^2) - 1.8(e^2 / \varepsilon R)$, where $E_g$ is the bulk material energy gap, μ is the reduced effective mass, e is the elementary charge, and ε is the dielectric constant. From this, a mean radius of about R~2.0 nm was estimated for the CdS QDs, thus confirming strong size quantum confinement **[16]**.

Furthermore, the increase in x-concentration clearly induced a blue shift in the OA band of the $Cd_{1-x}Mn_xS$ QDs from ~3.10 eV (x = 0.000) to ~3.22 eV for the highest magnetic doping (x = 0.100). Since these QDs were grown under identical synthesis conditions within the glass environment, it is expected that they would have the same mean size. As a result, there were no significant differences in the quantum confinements of these QDs that would cause shifts in the OA band peaks. Thus, it was concluded that the observed blue shift in OA band peak

(Fig. 1a) was a consequence of the **sp-d** exchange interactions between electrons confined in dot states and those located in the partially filled $Mn^{2+}$ states. This explanation is reasonable since replacing $Cd^{2+}$ with $Mn^{2+}$ ions should increase the energy gap of $Cd_{1-x}Mn_xS$ QDs**[18]**. In addition, it is interesting to note the weak **sp-d** exchange interaction in the $Cd_{1-x}Mn_xS$ bulk-like NCs because their OA band remains in an almost fixed position (~2.58 eV).

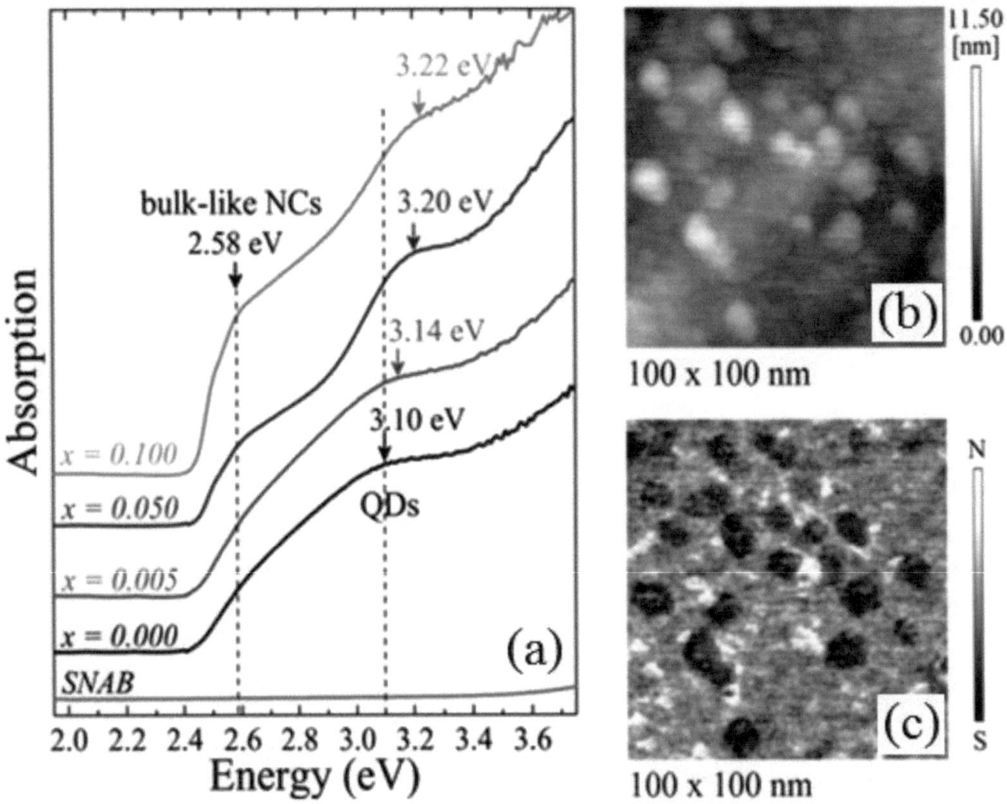


**Figure 1.** (a) Room temperature OA spectra of $Cd_{1-x}Mn_xS$ NPs with different x-concentrations embedded in the SNAB glass matrix. The two groups of NPs (QDs and bulk-like NCs) are indicated by the vertical dashed lines. The OA spectrum of the SNAB glass matrix is also shown at the bottom for comparison. (b) Topographic MFM image showing high quantities of $Cd_{0.900}Mn_{0.100}S$ NPs at the sample's surface, and (c) the corresponding phase MFM image (30 nm lift) where the contrast between the North (N) and South (S) magnetic poles identifies the orientation of the total magnetic moment of the DMS NPs.

Figure 1b presents the two-dimensional (100 x 100 nm) topographic MFM image of the sample with the highest level of magnetic doping (x = 0.100). Like the OA spectra, the topographic MFM image confirms the formation of two well defined groups of NPs with different mean radii: (i) R ~ 2.1 nm for the QDs, which closely agrees with the result estimated from the OA data (R ~ 2.0 nm); and (ii) R ~ 10.0 nm for the bulk-like NCs, a value near the vertical scale edge of Fig. 1b. Evidently, the exciton Bohr radius of bulk $Cd_{1-x}Mn_xS$ with diluted magnetic doping should be near that of bulk CdS, which is around $a_B$ ~ 3.1 nm **[16]**. Hence, we can conclude that the QDs with mean radius R ~ 2.0 nm are under strong quantum confinement, while the bulk-like NCs with mean radius R ~ 10.0 nm hardly exhibit any size confinement**[19]**.

In addition, a large quantity $Cd_{1-x}Mn_xS$ NPs can be observed in Fig. 1b, as well as in the corresponding phase MFM image shown in Fig. 1c. These images reveal great proximity between the two groups of NPs (QDs and bulk-like NCs), so that strong coupling between their wave functions is expected. In Fig. 1c, the topographic signal can be neglected because its phase MFM was recorded with a 30 nm lift from the sample's surface. Thus, interaction between tip and NP magnetization induces the contrast observed in this phase MFM image. The dark area (light area) is caused by attraction (repulsion) between tip and NP magnetization represented by the South (North) magnetic pole in the vertical scale bar of Fig. 1c. Evidently, the magnetization in each NP (QD or bulk-like NC) is caused by the size-dependent **sp-d** exchange interactions, proving that $Mn^{2+}$ ions are incorporated into the DMS nanostructures. This $Mn^{2+}$ ion incorporation in NPs has also been established by electron paramagnetic resonance (EPR) measurements and simulations with other samples synthesized in the same way as in this research **[17]**. In Fig. 1c, it is interesting to note that there is a relationship between the NP size and the direction of its magnetic moment: small (large) NPs have their magnetic moment oriented towards the North (South) pole.

Figures 2a and b present, as examples, the effect of temperature on $Cd_{1-x}Mn_xS$ NP luminescence with x = 0.000 and 0.050. The emissions from the two groups of $Cd_{0.950}Mn_{0.050}S$ NPs with different sizes, QDs and bulk-like NCs, are clearly identified in Fig. 2b by the presence of two well defined PL bands which are in agreement with the OA spectra of Fig. 1. However, in Fig. 2a, a PL band can be observed whose complex nature is a result of the overlapping of several emissions, including those from deep defects: denominated as (1) and (2) for the QDs, as well as $(1)^b$ and $(2)^b$ for the bulk-like NCs. In a recent study of other similar $Cd_{1-x}Mn_xS$ NPs with wurtzite structure, the existence of emissions from two trap levels related to the presence of deep defects was demonstrated**[20]**. The origin of these defects in $Cd_{1-x}Mn_xS$ NPs (and CdS NPs) with hexagonal wurtzite structure is possibly related to two energetically different $V_{Cd} - V_S$ divacancies: one oriented along the hexagonal c-axis (assigned to trap (1)), and the other oriented along the basal Cd-S bond (assigned to trap (2))**[20]**. Furthermore, the size-dependence of these trap-levels, (1) and (2), has been confirmed for CdSe NCs **[21]**, explaining the observed emissions from them in both the QDs ( $E_1$ and $E_2$ ) and bulk-like NCs ( $E_1^b$ and $E_2^b$ ) that are embedded in our glass samples.

In Figs. 2a and b, all emissions are marked by vertical dotted lines, including the bound exciton emission ($E_{exc}$) of QDs as well as the electron-hole recombination ($E_b$) of bulk-like NCs. The characteristic emission of $Mn^{2+}$ ions ($E_{Mn}$~2.12 eV) between the $^4T_1$ – $^6A_1$ levels in the $Cd_{1-x}Mn_xS$ NPs (with x ≠ 0) is also evident and represented in the Fig. 2c by $1/\tau_r^{Mn}$ rate **[1,22,23]**. The complete recombination aspects of these PL spectra are well-described in a diagram in Fig. 2c, where six (seven) emission bands can be identified for the CdS NPs ($Cd_{1-x}Mn_xS$ NPs with x ≠ 0). In Fig. 2b, the asymmetric shape of the emission band around 480 nm at low temperatures confirms the presence of shallow virtual levels for the QDs, and evidently there is also for the bulk-like NCs, as depicted in Fig. 2c. However, this emission band (480 nm) becomes symmetric with rising temperature, which demonstrates that the trapped carriers in the virtual levels are being released to other non-radiative channels of QDs. It is interesting to note that in Fig. 2a the excitonic emission ($E_{exc}$) of CdS QDs is almost suppressed due to the strong presence of non-radiative channels, including one related to the energy transfer from QDs to bulk-like NCs. However, a comparison between the PL spectra of the CdS and the $Cd_{0.950}Mn_{0.050}S$ NPs (see Fig. 2) clearly reveals that increasing x-concentration induces gradual suppression of emissions from all trap-levels ((1), (2), $(1)^b$, and $(2)^b$), since $Mn^{2+}$ ions are replacing the $V_{Cd}$ vacancies in the NPs. Indeed, this fascinating behavior provides further evidence that the deep defects are caused by $V_{Cd}$ –$V_S$ divacancies, and that the NPs are actually being doped by $Mn^{2+}$ ions. Hence, the non-radiative channels that supply the deep trap-levels disappear with increasing x-concentration in $Cd_{0.950}Mn_{0.050}S$ NPs, as shown in Fig. 2b.

In Fig. 2c, the wavy arrows represent non-radiative channels from the excitonic states of QDs, and from the conduction band (CB) of bulk-like NCs. Here, non-radiative energy transfer (ET) is given by the rate $1/\tau_{ET}^{n}$ (with n = A, B, C, A′, and B′), where $\tau_{ET}^{n}$ is the carrier escape time from an NP to one of these five non-radiative transitions. In our model, we have assumed that the non-radiative paths from the excitonic states of QDs, as well as from the conduction band of bulk-like NCs to the deep trap-levels ((1), (2), $(1)^b$, and $(2)^b$) can be disregarded. However, it is evident that these deep trap-levels may be filled by carriers from: (i) the shallow virtual levels of QDs and bulk-like NCs; and (ii) the $^4T_1$ levels of $Mn^{2+}$ ions**[20]**. Energy transfers from the excitonic states of QDs follow three paths: (A) to virtual levels (QDs); (B) to the conduction band of bulk-like NCs; and (C) to the $^4T_1$ level of $Mn^{2+}$ ions. On the other hand, the energy transfers from the conduction band of bulk-like NCs follow two paths: (A′) to virtual levels (bulk); and (B′) to the $^4T_1$ level of $Mn^{2+}$ ions. It is well known that the very fast energy transfer from a NP to $Mn^{2+}$ ions is generally resonant due to the high density of states above the emissive $^4T_1$ level,[1] as shown by the $^{2,4}\Gamma$ levels in Fig. 2c. However, size quantum confinement can play an important role in this process that, besides being mediated by the **sp-d** exchange interactions, is strongly dependent on the $Mn^{2+}$ fraction in $Cd_{1-x}Mn_xS$ NPs. In other words, QDs and bulk-like NCs are expected to behave differently due to the strong confinement of the QDs with a small mean radius of about R~2.0 nm.

The excitonic states of QDs can be denoted by $|1\rangle$, and the CB of bulk-like NCs by $|1^b\rangle$. The carrier number (depending on temperature T) of these two states is given by $N_1(T)$ and $N_1^b(T)$, respectively. Since carriers are thermally distributed each one of the three non-radiative channels related to QDs is supplied by $N_1(T)\exp(-E_n/K_BT)$ carriers, where $E_n$ (with n = A, B, and C) is the corresponding activation energy of the non-radiative n channel. Similarly, $N_1^b(T)\exp(-E_n/K_BT)$ carriers are transferred to each one of the two non-radiative channels related to bulk-like NCs, where n = A′, and B′. Furthermore, as shown in Fig. 2c by the straight, downward pointing arrows, radiative emissions are also present from both QDs and bulk-like NCs in the PL spectra which are related to $1/\tau_r^{QD}$ and $1/\tau_r^b$ rates, respectively. The straight, upward pointing arrow, indicated by g (g′), represents photo-excitation of the QDs (bulk-like NCs) caused by the laser pump. The carrier dynamics that take into account these transitions from the $|1\rangle$ (QD) and $|1^b\rangle$ (bulk-like NC) levels can be described by the following rate equations:

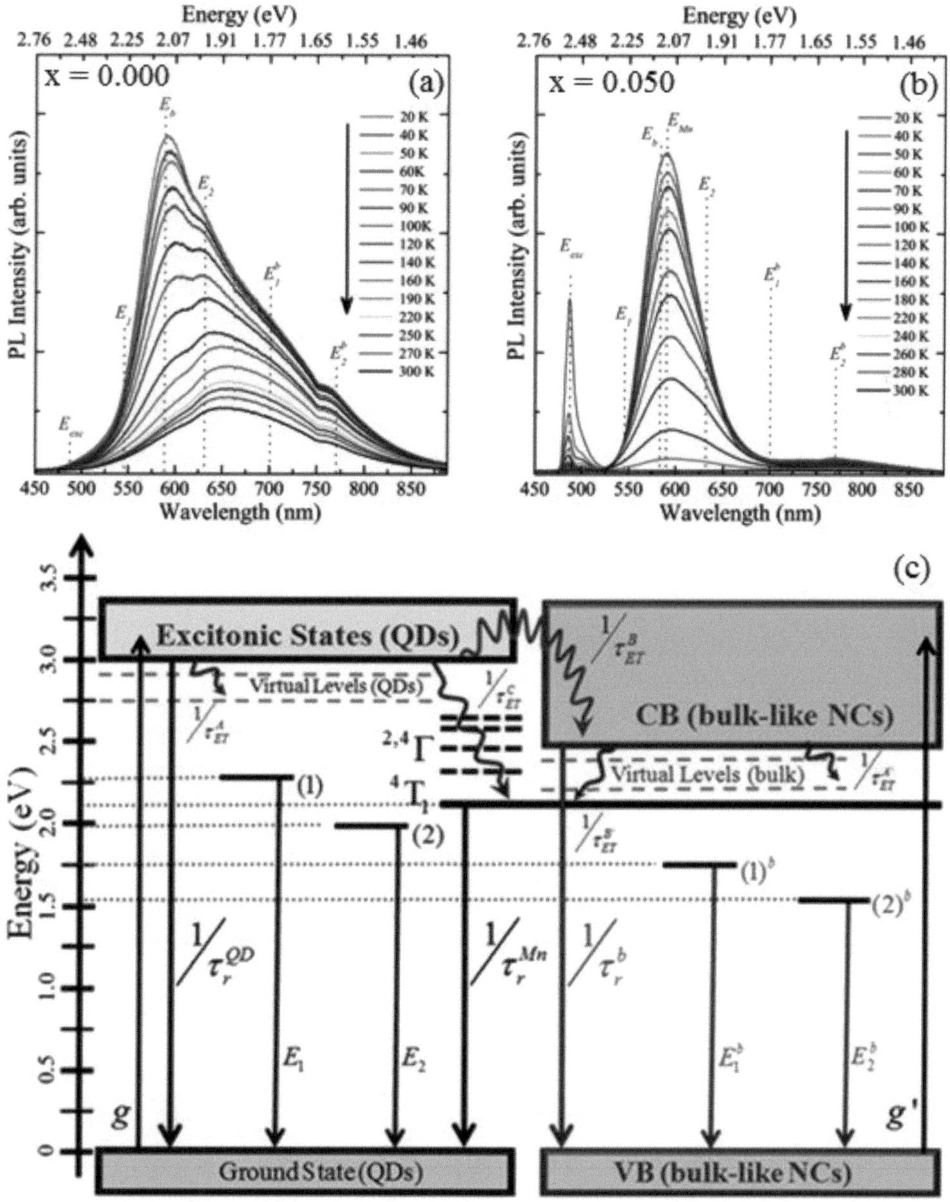


**Figure 2.** PL spectra of both (a) the CdS NPs (x = 0.000) and (b) $Cd_{0.950}Mn_{0.050}S$ NPs at several temperatures, from 20 K (top) to 300 K (bottom), as indicated by the downward pointing arrows. Their recombination aspects are depicted in panel (c), where the emissions from both the QDs and the bulk-like NCs are clearly identified. In addition, the characteristic emission of $Mn^{2+}$ ions ($^4T_1$–$^6A_1$), $E_{Mn}$ ~2.12 eV, when substitutionally incorporated in II-VI semiconductors is also evident. In the present energy scale, the $^6A_1$ level of the $Mn^{2+}$ ions is located at top of the QD ground state.

$$\frac{dN_1(T)}{dt} = +g - \underbrace{\frac{N_1(T)}{\tau_r^{QD}}}_{\text{radiative emission}} - \underbrace{\frac{N_1(T)\beta_A}{\tau_{ET}^{A}}}_{\text{QDs}\rightarrow\text{virtual levels (QDs)}} - \underbrace{\frac{N_1(T)\beta_B}{\tau_{ET}^{B}}}_{\text{QDs}\rightarrow\text{bulk-like NCs}} - \underbrace{\frac{N_1(T)\beta_C}{\tau_{ET}^{C}}}_{\text{QDs}\rightarrow Mn^{2+}\text{ ions}} ; \quad (1)$$

$$\frac{dN_1^b(T)}{dt} = +g' + \underbrace{\frac{N_1(T)\beta_B}{\tau_{ET}^{B}}}_{\text{QDs}\rightarrow\text{bulk-like NCs}} - \underbrace{\frac{N_1^b(T)}{\tau_r^{b}}}_{\text{radiative emission}} - \underbrace{\frac{N_1^b(T)\beta_{A'}}{\tau_{ET}^{A'}}}_{\text{bulk-like NCs}\rightarrow\text{virtual levels (bulk)}} - \underbrace{\frac{N_1^b(T)\beta_{B'}}{\tau_{ET}^{B'}}}_{\text{bulk-like NCs}\rightarrow Mn^{2+}\text{ ions}} ; \quad (2)$$

where $\beta_n = \exp(-E_n/K_BT)$ with n = A, B, C, A', and B'. In Eqs. (1) and (2), both the radiative emissions from QDs and bulk-like NCs and all non-radiative energy transfers are highlighted. In steady-state conditions, the laser excitations are given by $g = N_{1(0)}/\tau_r^{QD}$ and $g' = N_{1(0)}^b/\tau_r^b$ for QDs and bulk-like NCs, respectively. Moreover, there are no temporal changes in the carrier numbers, i.e., $\left(dN_1(T)/dt\right) = 0$ and $\left(dN_1^b(T)/dt\right) = 0$. When these conditions are replaced in Eqs. (1) and (2), we get:

$$N_1(T) = \frac{N_{1(0)}}{\left[1 + \alpha_A \exp\left(\frac{-E_A}{K_BT}\right) + \alpha_B \exp\left(\frac{-E_B}{K_BT}\right) + \alpha_C \exp\left(\frac{-E_C}{K_BT}\right)\right]} ; \quad (3)$$

$$0 = +\frac{N_{1(0)}^b}{\tau_r^b} + N_1(T)\frac{\exp\left(\frac{-E_B}{K_BT}\right)}{\tau_{ET}^{B}} - N_1^b(T)\left[\frac{1}{\tau_r^b} + \frac{\exp\left(\frac{-E_{A'}}{K_BT}\right)}{\tau_{ET}^{A'}} + \frac{\exp\left(\frac{-E_{B'}}{K_BT}\right)}{\tau_{ET}^{B'}}\right] . \quad (4)$$

The carriers' number in the QDs ($|1\rangle$ level) as a function of temperature T is given by Eq. (3), where the term $\alpha_n = \left(\tau_r^{QD}/\tau_{ET}^{n}\right)$ (with n = A, B, and C) can be considered constant at first approximation. After replacing the term $N_1(T)$ (given by Eq. (3)) in Eq. (4), we get:

$$0 = +\frac{N_{1(0)}^b}{\tau_r^b} + \frac{N_{1(0)}}{\tau_{ET}^{B}\left[\exp\left(\frac{E_B}{K_BT}\right) + \alpha_A \exp\left[\frac{-(E_A - E_B)}{K_BT}\right] + \alpha_B + \alpha_C \exp\left[\frac{-(E_C - E_B)}{K_BT}\right]\right]} - N_1^b(T)\left[\frac{1}{\tau_r^b} + \frac{\exp\left(\frac{-E_{A'}}{K_BT}\right)}{\tau_{ET}^{A'}} + \frac{\exp\left(\frac{-E_{B'}}{K_BT}\right)}{\tau_{ET}^{B'}}\right] . \quad (5)$$

Evidently, the second term on the right side of Eq. (5) is related to the carrier-mediated energy transfer from QDs to bulk-like NCs, and can be defined by:

$$\frac{N_{1(0)}^{b}(T)}{\tau_r^b}=\frac{N_{1(0)}}{\tau_{ET}^{B}\left[\exp\left(E_B/K_BT\right)+\alpha_A\exp\left[-(E_A-E_B)/K_BT\right]+\alpha_B+\alpha_C\exp\left[-(E_C-E_B)/K_BT\right]\right]}. \quad (6)$$

This represents temperature-dependent excitation of bulk-like NCs caused by carriers transferred from the QDs. Thus, Eq. (5) can be solved, resulting in the following expression:

$$N_1^b(T)=\frac{N_{1(0)}^b+N_{1(0)}^b(T)}{\left[1+\alpha_{A'}\exp\left(-E_{A'}/K_BT\right)+\alpha_{B'}\exp\left(-E_{B'}/K_BT\right)\right]}, \quad (7)$$

where $\alpha_n=\left(\tau_r^b/\tau_{ET}^n\right)$ with n = A′, and B′. Eq. (7) describes the temperature dependence for the carrier number of the bulk-like NCs ($|1^b\rangle$ level), and the term $N_{1(0)}^b(T)$ is given by Eq. (6). From Eqs. (3) and (7), we can find the steady-state intensities $I^{QD}(T)=\left[N_1(T)/\tau_r^{QD}\right]$ and $I^b(T)=\left[N_1^b(T)/\tau_r^b\right]$ of the QDs and bulk-like NCs, respectively, which results in:

$$I^{QD}(T)=\frac{I_0^{QD}}{\left[1+\alpha_A\exp\left(-E_A/K_BT\right)+\alpha_B\exp\left(-E_B/K_BT\right)+\alpha_C\exp\left(-E_C/K_BT\right)\right]}; \quad (8)$$

$$I^b(T)=\frac{I_0^b(T)}{\left[1+\alpha_{A'}\exp\left(-E_{A'}/K_BT\right)+\alpha_{B'}\exp\left(-E_{B'}/K_BT\right)\right]}. \quad (9)$$

In Eq. (8), it is interesting to note that the term $I_0^{QD}=\left[N_{1(0)}/\tau_r^{QD}\right]$ is temperature-independent because it is only related to QD photo-absorption. On the other hand, in Eq. (9), the term $I_0^b(T)=\left\{\left[N_{1(0)}^b(T)+N_{1(0)}^b\right]/\tau_r^b\right\}$ is temperature dependent and is given by Eq. (6) since the carrier-mediated energy transfer from QDs to bulk-like NCs is strongly temperature dependent. Evidently, there is coupling between Eqs. (8) and (9), and they can be fit to the experimental integrated PL intensity. This in turn, permits the deduction of activation energies related to the non-radiative channels of QDs ($E_A$, $E_B$, and $E_C$) as well as of bulk-like NCs ($E_{A'}$, and $E_{B'}$).

Figures 3a, b, and c show integrated PL intensity behavior for the doped $Cd_{1-x}Mn_xS$ NPs ($x \neq 0$) as a function of temperature. Here, the solid and open triangle symbols represent the bulk-like NCs and QDs, respectively. At low temperatures, QD emission intensity decreases quickly while bulk-like NC emissions remain almost constant except for a small increase at x = 0.100 (Fig. 3c). This behavior is due to the trapping of excited carriers from the excitonic states to the shallow virtual levels of QDs, where temperature increases induce a gradual release of these carriers to other electronic states, including the CB of bulk-like NCs. This carrier-mediated energy transfer from QDs to bulk-like NCs is a tunnelling phenomenon that is strongly dependent on the coupling between the wave functions of these NPs**[24]**. This effect is expected, given the high proximity between QDs and bulk-like NCs as confirmed by the MFM images (Figs. 1b and c). The ratio between these PL peak intensities (bulk-like NCs/QDs) as a function of temperature is shown in the insets of Fig.3. Here, it can be seen that the ratio increases at low temperatures and then decreases as QD emissions remain constant and bulk-like NC emissions decrease.

In the insets of Fig. 3, a fitting procedure with a Gaussian-like component, gives the temperature that yields the maximum ratio for each x-concentration: 122 K (x = 0.005); 134 K (x = 0.050); and 127 K (x = 0.100). Moreover, the FWHM (Full Width at Half Maximum) of the Gaussian-like component broadens with increasing x-concentration: 63 K (x = 0.005); 70 K (x = 0.050); and 74 K (x = 0.100), thus confirming that emission intensity from bulk-like NCs decreases more slowly after the maximum ratio is reached. It is interesting to note that the peak ratio between the PL intensities of bulk-like NCs/QDs is related to the inflection point of the corresponding integrated PL intensity of bulk-like NCs. This is indicated by the dashed vertical lines in Figs. 3a, b, and c. The inflection point temperatures were attributed to the maximum thermal energy transfer process from QDs to bulk-like NCs.

It can be seen that the temperatures obtained by the Gaussian fitting (T = 122 K, 134 K, and 127 K) can be related to delocalization thermal energies (like $K_BT$)**[25]**, which are needed to release the trapped carriers at shallow virtual levels (surface defects, for example) of QDs. Thus, the aforementioned $E_A$ activation energy coupled to these virtual levels (QDs) could be found by using the following expression: $E_A = K_B T$ ; where $K_B$ is the Boltzmann constant, and T is the temperature obtained by the Gaussian fitting. As a result, the x-concentration dependent behavior of this $E_A$ activation energy is given by: 10.51 meV (x = 0.005); 11.54 meV (x = 0.050); and 10.94 meV (x = 0.100), where the deduced values remain almost invariable. This result can take into account two effects caused by the increasing x-concentration of $Cd_{1-x}Mn_xS$ QDs: (i) the increasing energy gap that was observed in OA spectra of Fig. 1a; and (ii) possible density amplification of virtual levels associated to shallow defects of QDs**[23]**. Therefore, the combination of these effects in the electronic structure of $Cd_{1-x}Mn_xS$ QDs ($x \neq 0$) explains the nearly constant values obtained for the $E_A$ activation energy.

In order to deduce the additional activation energies ($E_B$ and $E_C$) related to other non-radiative channels of doped $Cd_{1-x}Mn_xS$ NPs ($x \neq 0$), Eqs. (8) and (9) were used to fit experimental integrated PL intensities as a function of reciprocal temperature (1/T), as shown in Figs. 4b, c, and d for the concentrations x = 0.005, 0.050, and 0.100, respectively. First, $E_B$ and $E_C$ activation energies related to QDs were determined by using Eq. (8) in which, with exception of the previously found $E_A$ activation energy, the following terms were used as parameters of fit: $I_0^{QD}$ , $E_B$, $E_C$ and $\alpha_n$ with n = A, B and C. Then, with the QD results, the activation energies related to bulk-like NCs ($E_{A'}$, and $E_{B'}$) could be found by fitting with Eq. (9).

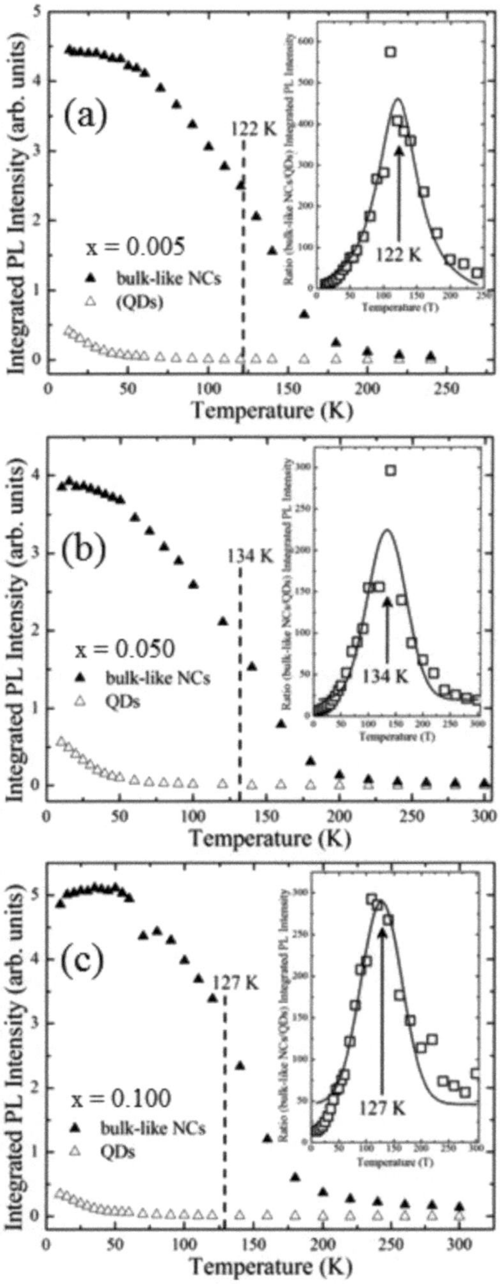


**Figure 3.** Temperature dependence of the integrated PL intensity of $Cd_{1-x}Mn_xS$ NPs at several x-concentrations: (a) x = 0.005; (b) x = 0.050; and (c) x = 0.100. QDs and bulk-like NCs are represented by open and solid triangle symbols, respectively. In the inset of each panel, the square symbols represent the ratio between these integrated PL intensities (bulk-like NCs/QDs), where fitting with a Gaussian-like component was used to find the temperature corresponding to the maximum value. The dashed vertical lines show that each one of these temperatures is close to the inflection point of the integrated PL intensity of the bulk-like NCs.

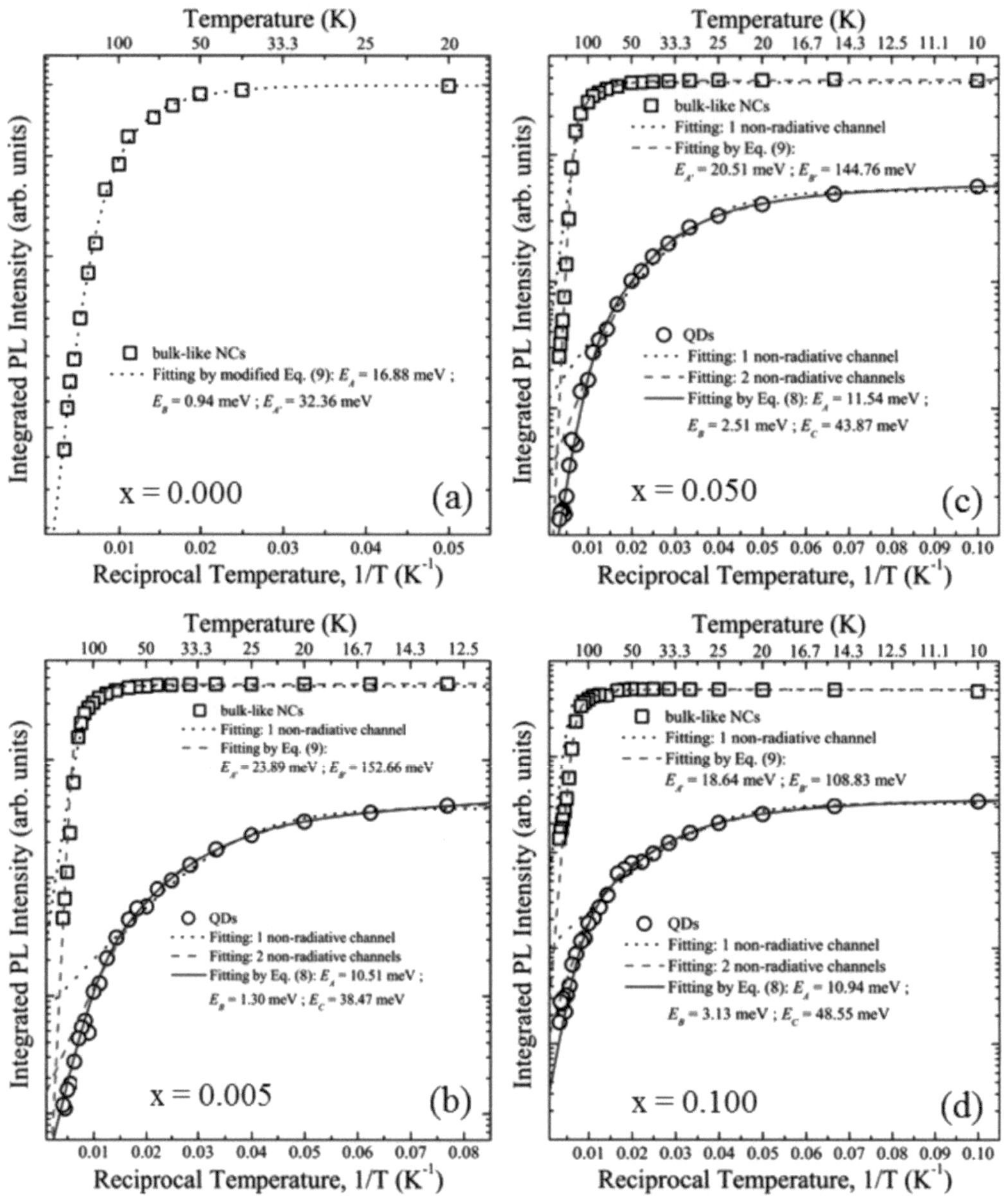


**Figure 4.** Experimental integrated PL intensity of $Cd_{1-x}Mn_xS$ NPs as a function of reciprocal temperature: (a) x = 0.000; (b) x = 0.005; (c) x = 0.050; and (d) x = 0.100. Each panel shows the fitting curves for the specified equations.

For the undoped NPs (x = 0), only a PL emission band associated with the bulk-like NCs could clearly be observed (see Fig. 2a). Therefore, it was not possible to use Eq. (8) to find the activation energies associated with the QDs. In addition, since there was no magnetic doping for these NPs (x = 0), it is expected that the non-radiative channels related to $Mn^{2+}$ ions would not exist. With these alterations, a modified Eq. (9) was used in fitting the experimental integrated PL intensity of CdS bulk-like NCs, where the $E_A$, $E_B$, and $E_{A'}$ activation energies were considered as parameters of fit. Figure 4a shows that good fit of the experimental data was achieved which confirms the absence of non-radiative channels related to $Mn^{2+}$ ions. Thus, even though any PL emissions from CdS QDs were not observed (see Fig. 2a), the $E_A$ and $E_B$ activation energies associated with them could be indirectly determined in this fitting procedure due to the carrier-mediated energy transfer from the QDs to the bulk-like NCs that is also present in the modified Eq. (9).

Furthermore, in Figs. 4b, c, and d, the fittings for both the QDs (Eq. (8)) and bulk-like NCs (Eq. (9)) are in excellent agreement with the experimental data. However, these concordances were not achieved by further fittings given: (i) one or two non-radiative channels for QDs ($x \neq 0$), and (ii) one non-radiative channel for bulk-like NCs ($x \neq 0$). Therefore, all these fittings evidently demonstrate that the Eqs. (8) and (9) for $Cd_{1-x}Mn_xS$ NPs ($x \neq 0$), as well as the modified Eq. (9) for CdS NPs, are satisfactorily suitable for describing the temperature-dependent carrier dynamics of the $|1\rangle$ or $|1^b\rangle$ levels.

Table 1 shows all the activation energies found that are related to non-radiative channels of $Cd_{1-x}Mn_xS$ NPs (QDs and bulk-like NCs) where, for doped NPs ($x \neq 0$), the $E_A$ remains almost constant as previously explained. Moreover, for undoped NPs the value $E_A \sim 16.88$ meV is slightly larger than that for doped NPs ($E_A \sim 11$ meV). This proves that increases in x-concentration enhance the density of the virtual levels associated with the shallow defects of QDs. The carrier-mediated energy transfer from QDs to bulk-like NCs, a tunnelling phenomenon, is evidently being hampered due to $E_B$ rising with increases in x-concentration (see Table I). In heavily doped NPs, there are many $Mn^{2+}$ ions incorporated near the surface of both groups of NPs (QDs and bulk-like NCs)**[22]**, an effect that enhances Mn–Mn interactions**[17,20]**. Therefore, we can conclude that high quantities of $Mn^{2+}$ ions near the surface of these NPs weakens the coupling between their wave functions which hampers the tunnelling process from the QDs to bulk-like NCs. Consequently, this effect also contributes to the excitonic emission ($E_{exc}$) of $Cd_{1-x}Mn_xS$ QDs, as observed in Fig. 2b.

| x-concentration | $E_A$ (meV) | $E_B$ (meV) | $E_C$ (meV) | $E_{A'}$ (meV) | $E_{B'}$ (meV) |
|---|---|---|---|---|---|
| 0.000 | 16.88 | 0.94 | ---- | 32.36 | ---- |
| 0.005 | 10.51 | 1.30 | 38.47 | 23.89 | 152.66 |
| 0.050 | 11.54 | 2.51 | 43.87 | 20.51 | 144.76 |
| 0.100 | 10.94 | 3.13 | 48.55 | 18.64 | 108.83 |

**Table 1.** Behavior of activation energies ($E_A$, $E_B$, $E_C$, $E_{A'}$, and $E_{B'}$) related to the non-radiative channels of $Cd_{1-x}Mn_xS$ NPs as a function of x-concentration. From the QDs, the non-radiative energy transfers are indexed as follows: ($E_A$) for virtual levels; ($E_B$) for the conduction band of the bulk-like NCs; and ($E_C$) for the $Mn^{2+}$ ions. For bulk-like NCs, the non-radiative energy transfers are denoted as: ($E_{A'}$) for the virtual levels; and ($E_{B'}$) for the $Mn^{2+}$ ions.

The non-radiative energy transfers from NPs to $Mn^{2+}$ions are related to the following activation energies: $E_C$ for the QDs; and $E_{B'}$ for the bulk-like NCs. In Table I, it can be seen that $E_C$ increases and $E_B$ decreases with rising x-concentration. This opposite behavior between QDs and bulk-like NCs demonstrates that the **sp-d** exchange interactions are strongly dependent on the size quantum confinement of the NPs. Increasing x-concentration from 0.000 to 0.100 induces considerable blue shift in the energy gap of the QDs, which can be disregarded for the bulk-like NCs (see Fig. 1a), while the density of the $^{2,4}\Gamma$ levels of $Mn^{2+}$ions is being amplified. Thus, the depth of $^{2,4}\Gamma$ levels is increasing in relation to the excitonic states of QDs, while remaining almost constant for the CB of bulk-like NCs. Therefore, the combination of these effects explains very well the observed increase (decrease) in $E_C$ ($E_{B'}$) activation energy with increasing x-concentration.

In addition, increasing x-concentration induces the density amplification of the virtual levels associated with the shallow defects of bulk-like NCs. This also occurs in the QDs**[26]**. However, since the change in energy gap of bulk-like NCs can be disregarded, these virtual levels become shallower for the conduction band (CB). Hence, in Table 1, the decrease in $E_{A'}$ activation energy with the increase in x-concentration can be adequately explained by taking into account this effect in the electronic structure of the bulk-like NCs.

### 3.2. Magneto-optical properties

Figure 5 shows the OA spectra, taken at room temperature, of $Cd_{1-x}Mn_xS$ magnetic NPs that were grown into the glass matrix environment. In Fig. 5a, the spectra were taken from NP samples with three different Mn-concentrations: x = 0.000, 0.050, and 0.100, and these samples did not have any thermal treatment. Only for comparison, the OA spectrum of the SNAB matrix is also shown in bottom of Fig. 5a and is clear the absence of any absorption band in the range between 350-650 nm. However, the OA spectra of all NP samples revealed the formation of two well defined groups of $Cd_{1-x}Mn_xS$ NPs with different sizes: (i) one group displaying a fixed band around 2.58 eV (near the energy gap of bulk CdS) and denominated as bulk-like NCs; (ii) the other displaying changing band energy due to quantum confinement properties and denominated as QDs.

A careful analysis of the bands attributed to $Cd_{1-x}Mn_xS$ QDs with concentrations x = 0.000; 0.050; and 0.100, clearly reveals a width of about 65 nm for each OA band which is due to a size distribution of the nanoparticles. From the OA peak at 3.13 eV, in Fig. 5a, and using the effective mass approximation **[12,17]**, an average radius around R~2.0 nm was estimated for CdS QDs (x = 0.000), which confirms the strong size quantum confinement. It is noted that an increase of Mn-concentration induces a blueshift on the QDs band, from 3.13 eV, for x = 0.000, to 3.22 eV, for x = 0.100. Since these magnetic QDs were synthesized under the same thermodynamic conditions, one should expect that they have the same average dot size (R~2.0 nm) and, thus, no significant differences in their quantum confinements that would cause shift among the OA band peaks. Here, can be also inferred that the growth kinetics of these dots is not influenced by the magnetic ions, since the amount of Mn dispersed in the glass environment is actually very small. Therefore, we attribute the blueshift on the peaks of Fig. 5a to the **sp-d** exchange interaction between electrons confined in the dot and located in the partially filled $Mn^{2+}$ ion states. This explanation is quite reasonable**[17]** since the replacement of Cd by Mn, in $Cd_{1-x}Mn_xS$ NPs, should change the energy gap between 2.58 eV, for CdS buk (x = 0) and 3.5 eV, for MnS bulk (x = 1).

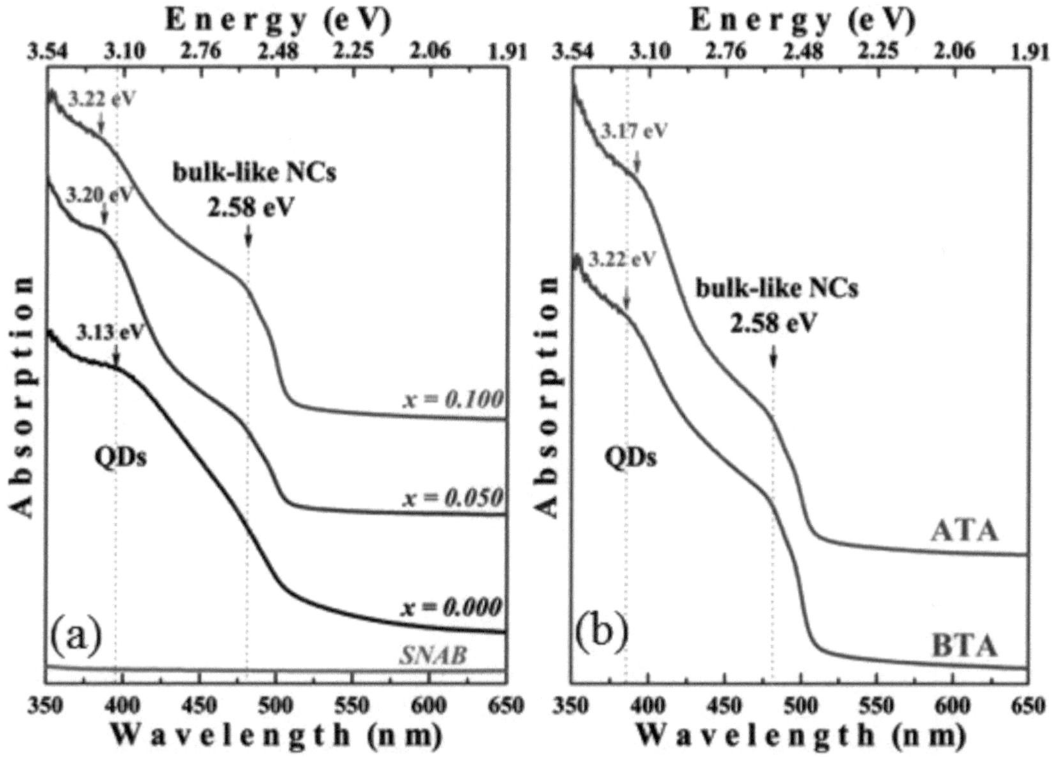


**Figure 5.** Room temperature OA spectra of $Cd_{1-x}Mn_xS$ NPs embedded in the SNAB matrix. Panel (a) shows the spectra of the as-grown samples which did not receive any thermal treatment. For comparison, it is also shown the OA spectrum of the SNAB matrix, at the bottom. Panel (b) shows spectra of two identical samples containing the same magnetic ion doping (x = 0.100), but one before thermal annealing (BTA) and the other after thermal annealing (ATA) at T = 560 ºC for 6 h. Observe that the peak at 2.58 eV, attributed to bulk-like NCs, does not change with doping or annealing.

We also compare the optical spectra of two identical $Cd_{0.900}Mn_{0.100}S$ samples, one before thermal annealing (BTA) and another after thermal annealing (ATA) at 560 ºC for 6h. The effect of this thermal annealing on the two OA band peaks of these NPs is shown in Fig. 5b. As expected for NPs without size quantum confinement, the OA band related to $Cd_{0.900}Mn_{0.100}S$ buk-like NCs does not show any shift in the samples with or without thermal annealing. At the same time, the $Cd_{0.900}Mn_{0.100}S$ QDs peak shows a redshift from 3.22 eV, in the sample without treatment (BTA), to ~3.17 eV, in the annealed sample (ATA). This shift can be ascribed to two possible annealing effects: (i) the size increase of the magnetic dot and thus, inducing weakening on the quantum confinement, and (ii) the decrease in the effective concentration of $Mn^{2+}$ ions incorporated into the dots during growth thus, inducing decreasing on the energy gap.

The dot size increase for increasing annealing time is a well known phenomenon governing the growth kinetic of nanoparticles in glass matrices**[18,27,28]**. However, a recent study of thermal treatments on undoped CdSe QDs**[12]** embedded in this same glass matrix (SNAB) showed a much smaller redshift (~0.03 eV) when annealed for 6 h. Since CdSe and CdS structures display great similarities, as well as $Cd_{1-x}Mn_xS$ with dilute Mn-concentration, it is reasonable to assume that they have the same growth kinetic in the same glass matrix. Thus, the higher shift (~0.05 eV) observed in $Cd_{0.900}Mn_{0.100}S$ QDs annealed for 6 h provides strong evidences that the observed higher redshift must also be ascribed to a decrease of the effective concentration of $Mn^{2+}$ incorporated to the dots, and this decrease takes place during the thermal treatment of the sample. We shall return to this evaporation-like process later, since its understanding is still an opened subject on doping processes in semiconductor QDs**[3]**.

Figure 6 presents the magnetic circularly polarized PL spectra, taken at 2.0 K and 15 T, of CdS (Fig. 6a) and of $Cd_{0.900}Mn_{0.100}S$ (Fig. 6b) NP samples. A decreasing from 300 K to temperatures near ~2.0 K causes an increase in the energy gap of CdS QDs, as well as in bulk CdS, of about ~ 85 meV**[29]**. Certainly, a similar temperature-dependent behaviour is expected for $Cd_{1-x}Mn_xS$ QDs with dot size R ~ 2 nm and for diluted magnetic doping. However, even with this large blueshift in the OA bands, the $Cd_{1-x}Mn_xS$ QDs with absorption around 405 nm, as well as all bulk-like NCs, were excited during the measurements at 2.0 K, due the large OA band width (~ 65 nm) of QDs as well as the wavelength width (± 5 nm) of the 405 nm excitation laser source. Therefore, according to OA spectra shown in Fig. 5, the emissions attributed to the two well-defined bulk-like NCs and QDs groups are observed in each spectrum of Fig. 6. The presence of virtual levels in these structures, as due to surface defects for example, can explain the asymmetric character of the emission band around 480 nm (Fig. 6b). Also, the broad emission band near 580 nm cannot be fitted by only one Gaussian-like component which provides further evidence for its complex nature associated to several emissions. We may conclude that besides the radiative recombination of excitons, labelled as $E_{QD}$ ($E_b$) for the QDs (bulk-like NCs), there are also the emissions from deep defect levels, labelled as (1) and (2) for QDs and $(1)_b$ and $(2)_b$ for bulk-like NCs. These emissions detected on the PL spectra are qualitatively described in the diagram depicted in Fig. 6c where the seven emission bands are identified. The deep defect levels in CdS and $Cd_{1-x}Mn_xS$ NPs with hexagonal wurtzite structure, a common phase for these materials,[30-32] are possibly related to two energetically different divacancy defects, $V_{Cd} - V_S$, associated to the absence of $Cd^{2+}$ and $S^{2-}$ ions in the crystalline NP structure[20]. One divacancy is oriented along the hexagonal c-axis of the wurtzite CdS structure and assigned to trap(1), whereas the other is oriented along the basal Cd-S bond directions and assigned to trap(2)[20,21]. The size dependence of these trapping levels has been confirmed for CdSe NCs[21], and is used to explain the detected emissions from QDs (labelled $E_1$ and $E_2$) and from bulk-like NCs (labelled $E_1^b$ and $E_2^b$) that occur in our $Cd_{1-x}Mn_xS$ structures, as depicted in Figs. 6.

The comparison between the emissions and absorptions of $Cd_{1-x}Mn_xS$ NPs, by taking into account the mentioned increase in energy gap at low temperature (~ 2.0 K), revels that these nanostructures embedded in a glass matrix exhibit an anomalously large Stokes shift ($\Delta_{SS}$) given by: $\Delta_{SS}$ ~ 0.65 eV for QDs, and $\Delta_{SS}$ ~ 0.54 eV for bulk-like NCs. Possibly, the origin for this large Stokes shift can be attributed to radiative recombination due the many-body effects on the excitonic states of the NPs, a phenomenon that was recently demonstrated for PbS nanocrystals**[21]** and should be considered for the $Cd_{1-x}Mn_xS$ NPs. Certainly, further investigations are required in order to reach a comprehensive explanation for these observed large Stokes shifts in the $Cd_{1-x}Mn_xS$ NPs embedded in a glass matrix.

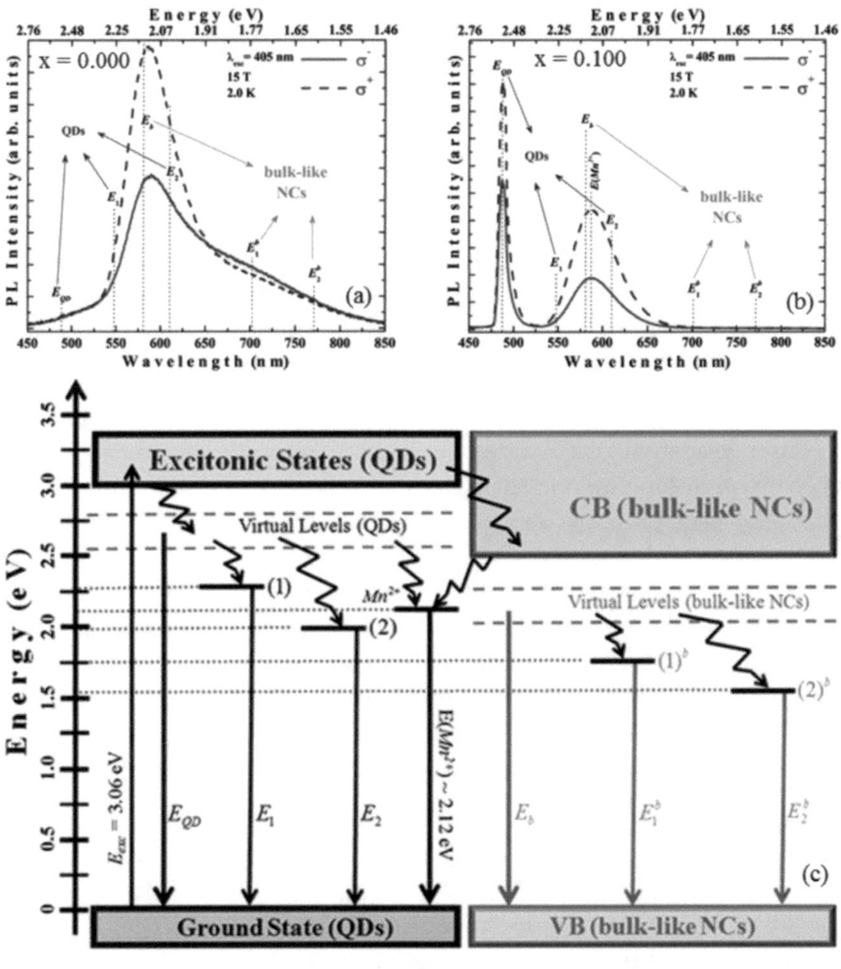


**Figure 6.** Circularly polarized PL spectra, $\sigma^-$ (solid lines) and $\sigma^+$ (dashed lines), taken at 2.0 K and magnetic field B = 15 T, are shown in panel (a) for undoped CdS NP sample and in panel (b) for magnetic $Cd_{0.900}Mn_{0.100}S$ NP sample. The different recombination processes are depicted in panel (c), where the emissions from QDs and from bulk-like NCs are clearly identified. The characteristic emission $E(Mn^{2+})$ of $Mn^{2+}$ ions ($^4T_1 \rightarrow {}^6A_1$), occurring near 2.12 eV when incorporated in the II-VI semiconductors, is almost resonant with the $E_b$ emission from bulk-like NCs. The non-radiative processes associated to the $V_{Cd} - V_S$ divacancies occurring in the structures are also indicated.

The characteristic emissions $^4T_1 \rightarrow {}^6A_1$ between levels of $Mn^{2+}$ ions, labelled as $E(Mn^{2+})$ in Figs. 6b and 6c and with transition energy ~2.12 eV, also confirm that these magnetic impurities were substitutionally incorporated in the $Cd_{1-x}Mn_xS$ NPs**[1,22,23]**. This incorporation of $Mn^{2+}$ ions in NPs has also been proved by electron paramagnetic resonance (EPR) measurements and simulations in other samples which were synthesized by the same method used in this work**[17,20]**.

Note that the emissions from deep defect levels observed in bulk-like NCs, and shown in Figs. 6a and 6c with labels $E_1^b$ and $E_2^b$, become almost suppressed in samples with magnetic doping (see Fig. 6b). Thus, $Mn^{2+}$ ions must be filling out the Cd vacancies, $V_{Cd}$, during doping and this interesting fact provides further evidence not only for the existence of the deep divacancies, $V_{Cd} - V_S$, but also for the incorporation of $Mn^{2+}$ ions in the NPs.

Figures 7a and 7b present the circularly polarized ($\sigma^-$ and $\sigma^+$) PL spectra of $Cd_{0.900}Mn_{0.100}S$ NP samples without thermal treatment and taken at 2.0 K, for several magnetic field values between 0.0 and 15.0 T. The magnetic subcomponent emissions from QDs and from bulk-like NCs can be clearly observed in all PL spectra. It is also noted that $\sigma^+$ increase the intensity faster than $\sigma^-$ emissions, thus resulting in the strong PL circular polarization. The relative intensity ratio between the polarized emissions from QDs and from bulk-like NCs is shown in Fig. 7c as a function of magnetic field for two different Mn-concentrations (x = 0.050 and 0.100).

The internal optical transition ($^4T_1 \rightarrow {}^6A_1$) occurring within excited $3d^5$ shells of the $Mn^{2+}$ ions is highly sensitive to the presence of external magnetic fields**[33,34]**. After electron-hole pair creation by laser excitation, the band-edge exciton can either recombine radiatively or transfer its energy to a $Mn^{2+}$ ion via an Auger-like process that depends on the exciton-Mn coupling. At low temperatures and in magnetic fields, this $Mn^{2+}$ PL band remains unpolarized and, eventually, becomes suppressed while the circularly polarized band-edge excitonic emissions increase the intensity. This is a universal behaviour that has been observed in DMS crystals, epilayers, quantum wells, quantum wires, and in self-assembled epitaxial quantum dots**[33-42]**. Although the precise mechanism of energy transfer from excitons to electrons in the $Mn^{2+}$ $3d^5$ shell is still debated in the scientific community,**[37,38,42-44]** the marked field dependence of this process indicates a spin-dependent excitation transfer as described by Nawrocki**[45]** and by Chernenko**[42,43]**.

The energy transfer from QDs to $Mn^{2+}$ ions is also highly sensitive to the presence of external magnetic fields**[2]**. For example, in self-organized $Cd_{1-x}Mn_xSe$ QD samples, the $^4T_1 \rightarrow {}^6A_1$ emissions from the incorporated $Mn^{2+}$ ions are completely suppressed at a magnetic field values near 3–4 T **[40]**. According to Nawrocki model**[45]**, this occurs because the $Mn^{2+}$ magnetization freezes out the electron population in the $M_s = -5/2$ ground state Zeeman sublevels of the $Mn^{2+}$ ions (labelled $^6A_1$) **[2,42-45]**. In this model, the transition of an electron from the conduction to the valence band occurs without change of its spin, and the transition is allowed if the total spin of the combined system of Mn-ion+electron is conserved. In particular, no Auger recombination is possible with participation of Mn-ion ground state ($^6A_1$) with spin S = 5/2 and $S_z = \pm 5/2$ since the excited state ($^4T_1$) has spin S = 3/2 and $S_z = \pm 3/2$. The suppression of the Auger recombination in the high magnetic field can be explained by assuming thermalization of Mn-ions in the lowest state with $S_z = -5/2$. Thus, the band-edge excitonic emission intensity saturates with increasing magnetic field, indicating alignment of the QD exciton spins by the magnetic field.

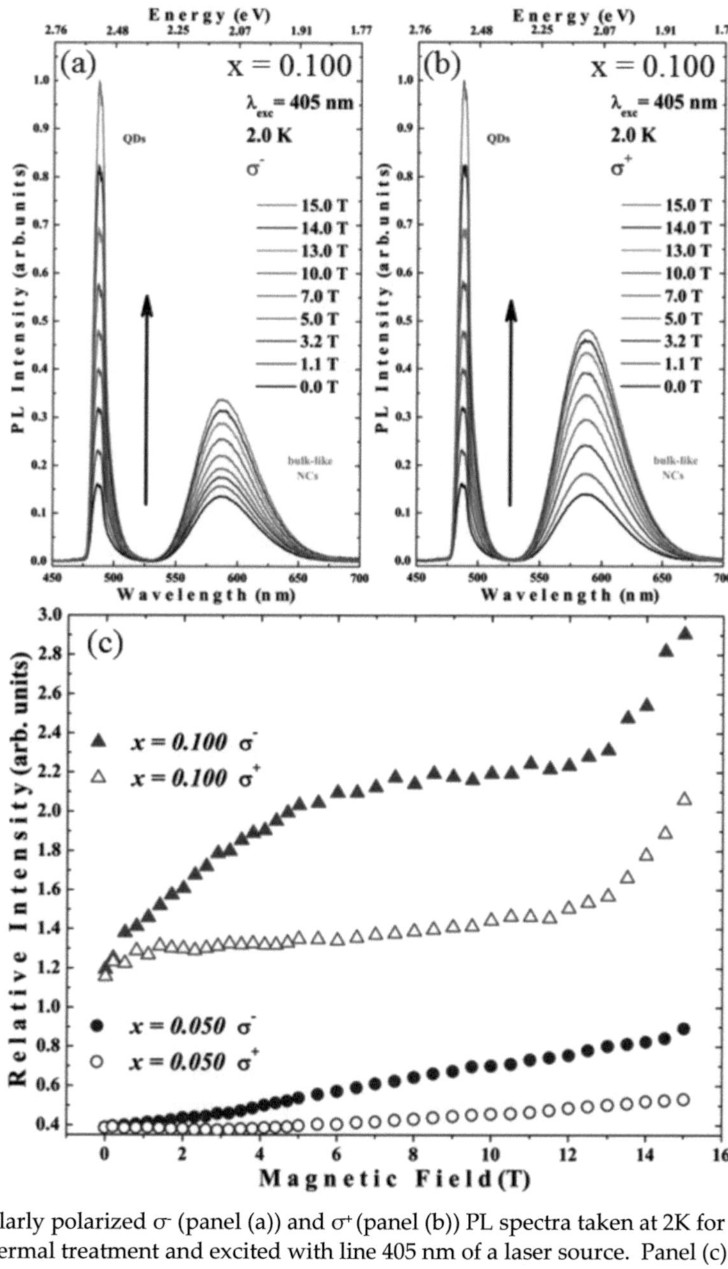


**Figure 7.** Circularly polarized $\sigma^-$ (panel (a)) and $\sigma^+$ (panel (b)) PL spectra taken at 2K for $Cd_{0.900}Mn_{0.100}S$ NPs without thermal treatment and excited with line 405 nm of a laser source. Panel (c): Comparison between the ratio of $\sigma^-$ (filled symbols) and $\sigma^+$ (opened symbols) emission intensities from QDs and from bulk-like NCs in samples with concentrations x = 0.050 (circles) and x = 0.100 (triangles) for increasing magnetic field values. Notice that magnetic doping affects strongly the magnetic dependence of these intensities.

Hence, in our samples is also expected that the electron-hole radiative recombinations ($E_{QD}$ and $E_b$) show increasing intensities while the E($Mn^{2+}$) emissions show decreasing change of intensities for increasing magnetic field. Because the overlap with the $E_b$ emissions in these samples, the E($Mn^{2+}$) emissions cannot be clearly resolved in the PL spectra (Figs. 7a and b) but should show a change in the relative intensity (Fig. 7c). For sample with Mn-concentration x = 0.100, the ratio between polarized PL intensities from QDs and from bulk-like NCs (triangles) displays non-monotonic behaviour from 0 to 12 T with a saturation tendency, where the increased exciton emissions increase occurring at the expenses of Mn PL emissions decrease, as the magnetic fields increase. In the sample with smaller concentration, x = 0.050, the ratio between PL intensities (circles) increases almost linearly up to 15 T, or even with slight intensity change. In Fig. 7c, it is also noted a significant change in the relative intensities of emissions for $\sigma$ polarization, starting at B ~ 4 T, in the sample with higher Mn-concentration (x = 0.100) and a much lower intensity change in the sample with Mn-concentration x = 0.050. It is our understanding that this effect is related to the suppressed E($Mn^{2+}$) emissions.

Furthermore, in Fig. 7c, the x-concentration dependent behaviour of the exciton intensity variation with the magnetic field indicates a strong modification of the Auger energy transfer rate from the excitons to $Mn^{2+}$ ions. Therefore, in the low Mn-concentration this energy transfer does not occur as strongly as in the case of the high Mn-concentration. It has been shown that Auger energy transfer is sensitively dependent on carrier density – excitation power**[40,42]** and Mn-concentration **[36]**. In the high power excitation, the suppression of the Auger process does not take place as strongly as in the case of the weak power excitation even in the high magnetic field region. In the low excitation intensity the PL intensity curve behaves very different as in the high excitation intensity. Based on the approach of Nawrocki et al.**[45]**, Chernenko et al.**[42,43]** calculated the increase in the exciton intensity with B and showed that this increasing is associated with lifetime of non-radiative transition $I(B)/I(0) = const/\left(1+\left(\tau_0/\tau_A\right)\right)$ were $\tau_0$ and $\tau_A$ are the times of radiative and non-radiative recombinations of the exciton, respectively. Here, the effective time of non-radiative recombination depends on B, Mn-concentration and carrier density **[36,42-44]**. Similar behaviour of the dependence of the PL intensity with B is observed for our samples with different Mn-concentrations. Note that the emissions from deep defect levels observed in bulk-like NCs, and shown in Figs. 6a and 6c with labels $E_1^b$ and $E_2^b$, become almost suppressed in samples with magnetic doping (see Fig. 6b). Thus, $Mn^{2+}$ ions must be filling out the Cd vacancies, $V_{Cd}$, during doping and this interesting fact provides evidence that $Mn^{2+}$ doping can alter the carrier density in the NPs.

The Zeeman energy splitting in the electronic structure of the NPs is also other important effect caused by the increasing in the magnetic field. It is well known that in DMS structures the Zeeman energy splitting can be considerably altered by the exchange interaction between the carrier spins and the substitutional doping magnetic ions**[46]**. Thus, in our $Cd_{1-x}Mn_xS$ NP samples, it is quite expected different Zeeman energy splitting for QDs and bulk-like NCs with the increase in the magnetic field. This in turn should also increase the separation between the excited electronic levels of QDs and bulk-like NCs, so that the non-radiative energy transfer between them (as depicted in Fig. 6b) is being weakened up to be completely interrupted at a given magnetic field value. We understand that this phenomenon occurs at a magnetic field B ~ 12 T for the $Cd_{1-x}Mn_xS$ NP sample with x = 0.100, contributing thus to the abrupt increase in relative PL intensity (QDs/bulk-like NCs) shown in Fig. 7c. Since this effect cannot be clearly observed for the $Cd_{1-x}Mn_xS$ NP samples with x = 0.050, we can conclude that non-radiative energy transfer involving the excited electronic levels of QDs and bulk-like NCs was not completely broken off in the investigated magnetic field range, favouring the almost linear behaviour observed in Fig. 7c. It is important to mention that the energy transfers involving the excitonic states of QDs, the conduction band of bulk-like NCs, and the shallow virtual levels of NPs was demonstrated in section 3.1 (carrier dynamics)**[47]**.

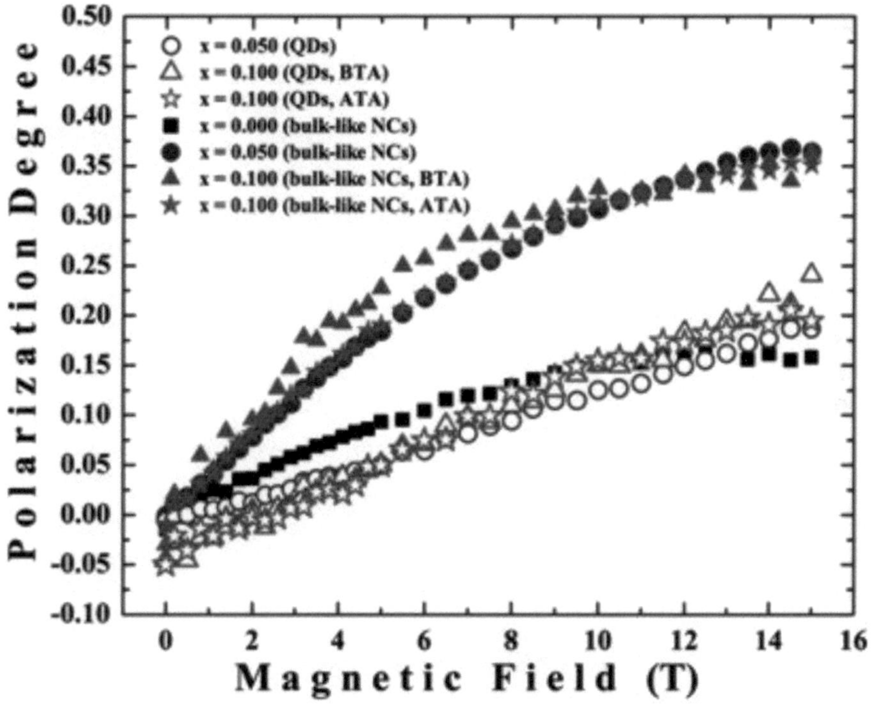


**Figure 8.** Magnetic field dependence of the polarization degree ( $\rho(B)$ ) of $Cd_{1-x}Mn_xS$ NPs: QDs (open symbols) and bulk-like NCs (filled symbols). For a comparison it is shown the degrees of polarization for two identical samples with concentrations x = 0.100, but one before thermal annealing (BTA) and another after undergoing thermal annealing (ATA) at T = 560 ºC for 6 h.

The degree of polarization is defined by $\rho(B) = (I_+ - I_-)/(I_+ + I_-)$ **[2,41,46],** where $I_+$ and $I_-$ are the integrated intensities of $\sigma^+$ and $\sigma^-$ magnetic circularly polarized PL (MCPL) spectra taken at a given magnetic field, B. The values of $\rho(B)$ for bulk-like NCs and for QD emissions are represented by filled and opened symbols in Fig. 8, respectively. For QD emissions, $\rho(B)$ increases almost linearly up to B = 15 T, and reaches 25% polarization. The bulk-like NC emissions appear to increase quadratically with B and, at higher magnetic fields, show a saturation tendency near 35%. However, the degree of polarization for CdS bulk-like NCs (x = 0.000) shows a much slower increase with saturation value near 15%. In the Mn-doped samples (x ≠ 0), the bulk-like NCs exhibit a higher degree of polarization than the QDs, thus evidencing that the amount of $Mn^{2+}$ ions that are substitutionally incorporated into QDs is smaller than in the bulk-like NC. Evidently, this effect is related to the well known difficulty in doping semiconductor QDs with magnetic impurities. Furthermore, the mentioned non-radiative energy transfer from QDs to bulk-like NCs is also a cause for the lower degree of polarization for the QDs. It is fascinating to note that the degree of polarization for the $Cd_{1-x}Mn_xS$ QDs with x = 0.100 becomes higher than for the undoped bulk-like NCs (x = 0) at the same magnetic field in which the abrupt increase in relative PL intensity (Fig. 7c) takes place, i. e., B ~ 12 T. For the $Cd_{1-x}Mn_xS$ QDs with x = 0.050 this effect is less pronounced and occurs at a higher magnetic field (see Fig. 8). In addition, the different magneto-optical properties of the bulk-like NCs and QDs can be explained by taking into account a considerable change of exchange interaction between the carrier spins and the substitutional doping of magnetic ions incorporated into the NPs with different sizes. This fact confirms that the size quantum confinement plays important role on the magneto-optical properties of $Cd_{1-x}Mn_xS$ NPs. We note that the Mn-doped with x = 0.050 and undoped CdS NP samples does not exhibit any zero-field degree of polarization. However, the samples doped with x = 0.100 presented negative zero-field polarization, $\rho(B=0) \cong -5\%$, which is ascribed to a change in the Zeeman ground state character. This interesting phenomenon is related to an intrinsic magnetism of $Cd_{1-x}Mn_xS$ NPs caused by the change in the **sp-d** exchange interaction strength, which is strongly dependent on the doping mole fraction x of incorporated magnetic ions.

A comparison between the degrees of polarization for two sets of identical Mn-doped samples with x = 0.100, is shown in Fig. 8, one before thermal annealing, labelled BTA and represented by triangles symbols; another undergone a thermal annealing at T = 560 ºC for 6 h, labelled ATA and represented by star symbols. After thermal annealing, the $Cd_{0.900}Mn_{0.100}S$ bulk-like NCs sample showed a degree of polarization very similar to the $Cd_{0.950}Mn_{0.050}S$ bulk-like NCs sample. This fact indicates that during the thermal annealing at T = 560 ºC occurred a decrease in the effective concentration ($x_{eff}$) of the $Cd_{1-x}Mn_xS$ bulk-like NCs. In agreement with observed redshift for OA band shown in Fig. 5b, this same dynamical doping process should also occur in the $Cd_{1-x}Mn_xS$ QDs. However, as shown in Fig. 8 (see open triangle and star symbols), the change in degree of polarization for QDs induced by thermal annealing is small due to two effects: (i) the strong localization of magnetic $Mn^{2+}$ ions in the same place as the charged carries confined to the dots, and (ii) the mentioned smaller amount of magnetic impurity that is incorporated into QDs.

Figures 9 (a, b, and c) presents two-dimensional phase MFM images (room temperature) of the $Cd_{1-x}Mn_xS$ NPs (x = 0.100) that are located at the samples surface, where it is possible to investigate the thermal annealing effect on the total magnetic moment of NPs. The images (150 x 150 nm) with a lift of 20 nm were recorded in two situations: in Fig. 9a before thermal annealing (BTA); and in Fig. 9c after thermal annealing (ATA). The contrast between these MFM images is a result of the interactions between the tip and the NP magnetization. However, there is also a small influence of the sample topography because the probe is close to the sample surface (20 nm). As a result of magnetic interaction between tip and surface, the bright area (dark area) of the phase MFM image displays repulsive (attractive) interaction. Thus, the clear contrast that is observed in Fig. 9a, which is mainly caused by magnetic interactions with the NPs, is almost vanished after thermal annealing as shown in Fig. 5c. Hence, we may conclude that the total magnetic moment of each NP (observed in Fig. 9a) is caused by the **sp-d** exchange interactions and can be tuned by a suitable thermal annealing of the $Cd_{1-x}Mn_xS$ NP samples. In agreement with the results of Figs. 5 and 8, this behaviour can ascribed to diffusion of $Mn^{2+}$ ions from the core to a position near the NP surface and, therefore, decreasing the effective concentration ($x_{eff}$) in $Cd_{1-x}Mn_xS$ samples. In addition, Figure 9b shows the phase MFM image (30 x 30 nm) obtained with no lift, of a $Cd_{1-x}Mn_xS$ NP (BTA) that is at the sample surface. Since this image was recorded with no lift, there is a strong influence of the sample topography and allowing the observation of the characteristic hexagon of the wurtzite structure.

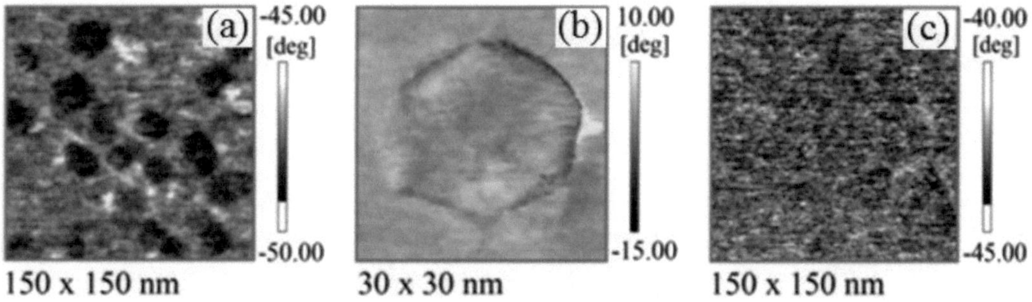


**Figure 9.** Room temperature phase MFM images (150 x 150 nm) with a lift of 20 nm of two $Cd_{0.900}Mn_{0.100}S$ samples: (a) before thermal annealing (BTA); and (C) after thermal annealing (ATA). Panel (B): Phase MFM image (30 x 30 nm) with no lift of a NP before thermal annealing, where the characteristic hexagon of the wurtzite structure can be observed.

The main doping models for QDs that are used to explain the incorporation of impurities, including $Mn^{2+}$ as in our samples, are known as 'trapped-dopant' and 'self-purification' mechanisms, which have being largely discussed in last few years**[3,48-52]**. The trapped-dopant mechanism is governed by the growth kinetics, where the impurity is adsorbed on the dot surface and then covered by additional material,**[3]** while the self-purification mechanism is governed by a diffusion process of impurities to more stable and stronger binding energy sites near the surface of dots**[48]**.

It becomes clear that the trapped-dopant mechanism is occurring in the course of the thermal annealing at T = 560 ºC of our sample, since the $Cd_{1-x}Mn_xS$ QDs are growing due to increasing annealing time (see Fig. 5b). However, it is necessary to answer the question: Is this mechanism responsible for the decreasing x-concentration of $Mn^{2+}$ ions in the QDs? In our conception, it is reasonable to assume that the trapped-dopant mechanism does not account for a significant change of Mn-concentration during the $Cd_{1-x}Mn_xS$ dot growth by the melting-nucleation synthesis. Since there is a relative homogeneous distribution of $Cd^{2+}$, $Mn^{2+}$ and $S^{2-}$ species into the glass environment in each moment of the thermal annealing, it is expected to observe nearly constant Mn-concentration in the $Cd_{1-x}Mn_xS$ dot growth process. Furthermore, the trapped-dopant mechanism is generally dominant for QD synthesis based on liquid phase approach, as the colloidal chemistry, where the temperatures are generally below 350 ºC and, in some case, even as low as room temperature**[3,53]**. In contrast, the energetic argument related to the self-purification mechanism imposes a relative instability for the impurity species due to increasing formation energy for decreasing dot-size. In Mn-doped CdSe NCs, for example, it is known that the diffusion of $Mn^{2+}$ ions occurs at a synthesis temperature around ~550 K (277 ºC), due to this instability **[54,55]**. Therefore, we are convinced that the relatively high temperature used in thermal annealing of our sample (560 ºC) is able to provide enough energy to provoke impurity diffusion toward surface region, a site having stronger binding energy, or even to evaporate the magnetic impurity ions from the $Cd_{1-x}Mn_xS$ QDs. In other words, our results confirm that self-purification is the dominant mechanism that controls the doping in semiconductor QDs grown by melting-nucleation synthesis approach.

## 4. Conclusions

In conclusion, we have recorded optical absorption (OA), photoluminescence (PL), and magnetic circularly polarized photoluminescence (MCPL) spectra, as well as magnetic force microscopy (MFM) images, in order to investigate $Cd_{1-x}Mn_xS$ NPs that were synthesized in a glass matrix. Room temperature OA spectra revealed the growth of two groups of NPs with different sizes: QDs and bulk-like NCs, a result confirmed by MFM images. Several emissions were observed in the temperature dependent PL spectra of $Cd_{1-x}Mn_xS$ NPs, including those from deep defect levels that were attributed to two energetically different divacancies, $V_{Cd}$–$V_S$, in the wurtzite structure. Moreover, the emissions from these deep defect levels were suppressed with increasing x-concentration, providing further evidence not only of the incorporation of $Mn^{2+}$ ions in the NPs, but also for the existence of deep divacancy defects $V_{Cd}$ – $V_S$. Therefore, we have demonstrated that the density of NP defects can be controlled by magnetic doping. From the temperature dependent PL spectra of these NPs, we have deduced, based on rate equation, expressions in order to describe the carrier dynamics between excitonic states of QDs and conduction band of bulk-like NCs. Fitting procedures with these coupled expressions achieved satisfactory agreement with the integrated PL intensity of both the QDs and bulk-like NCs provided activation energies of non-radiative channels observed in $Cd_{1-x}Mn_xS$ NPs.

Our results confirm that the magnetic doping, $Mn^{2+}$ ions localization, and quantum confinement play important roles on the magneto-optical properties of these NPs. The different behaviour observed between the two groups of NPs with different sizes, QDs and bulk-like NCs, were ascribed to a considerable change of exchange interaction between the carrier spins and the substitutional doping magnetic ions incorporated into the NPs. In addition, we have demonstrated that the relatively high temperature that was used in the thermal annealing of the samples provides enough energy to provoke magnetic impurity diffusion toward surface region of NPs. Therefore, for semiconductor QDs grown by the melting-nucleation synthesis approach, the doping process is dominated by the self-purification mechanism. We believe that the main results of this chapter can motivate further investigations and applications of other systems containing DMS NPs.

## Author details

Noelio Oliveira Dantas and Ernesto Soares de Freitas Neto
*Laboratório de Novos Materiais Isolantes e Semicondutores (LNMIS), Instituto de Física, Universidade Federal de Uberlândia,Uberlândia, Minas Gerais, Brazil*

## Acknowledgement

The authors gratefully acknowledge financial support from the Brazilian Agencies FAPEMIG, MCT/CNPq, and CAPES. We are also thankful for use of the facilities for the MFM measurements at the Institute of Physics (INFIS), Federal University of Uberlandia (UFU), supported by a grant (Pró-Equipamentos) from the Brazilian Agency CAPES. We are also grateful to our collaborators: Sidney A. Lourenço, Márcio D. Teodoro and Gilmar E. Marques.

## 5. References

[1] Archer P I, Santangelo S A, Gamelin D R (2007). Nano Lett. 7: 1037.

[2] Beaulac R, Archer P I, Ochsenbein S T, Gamelin D R (2008). Adv.Funct.Mater. 18: 3873.

[3] Norris D J, Efros A L, Erwin S C (2008).Science. 319: 1776.

[4] Gur I, Fromer N A, Geier M L, Alivisatos A P (2005).Science. 310: 462.

[5] Erwin S C, Zu L , Haftel M L, Efros A L, Kennedy T A, Norris D J (2005). Nature. 436: 91.

[6] Yu J H, Liu X, Kweon K E, Joo J, Park J, Ko K –T, Lee D W,Shen S, Tivakornsasithorn K, Son J S, Park J –H, Kim Y –W, Hwang G S, Dobrowolska M, Furdyna J K, Hyeon T (2010).Nature Materials. 9: 47.

[7] Beaulac R, Archer P I, Liu X, Lee S, Salley G M, Dobrowolska M, Furdyna J K, Gamelin D R (2008).Nano Lett. 8: 1197.

[8] Bacher G, Schweizer H, Kovac J, Forchel A, Nickel H, Schlapp W, Lösch R (1991). Phys Rev B. 43: 9312.
[9] Felici M, Polimeni A, Miriametro A, Capizzi M, Xin H P, Tu C W (2005). Phys Rev B. 71:045209.
[10] Sanguinetti S, Henini M, Alessi M Grassi, Capizzi M, Frigeri P, Franchi S (1999). Phys Ver B. 60: 8276.
[11] Banyai L, Koch S W (1993) Semiconductor Quantum Dots. Singapore: World Scientific Publishing Co.
[12] Freitas Neto E S, Dantas N O, da Silva S W, Morais P C, Pereira-da-Silva M A (2010). J Raman Spectrosc. 41: 1302.
[13] Beaulac R, Archer P I, Ochsenbein S T, Gamelin D R (2008). Adv Funct Mater. 18: 3873.
[14] Mocatta D, Cohen G, Schattner J, Millo O, Rabani E, Banin U (2011). Science. 332: 77.
[15] Brus L E (1984).J Chem Phys. 80: 4403.
[16] Grahn H T (1999) Introduction to Semiconductor Physics, World Scientific Publishing Co. Pte. Ltd., New York. Ch 9.
[17] Dantas N O, F Neto E S, Silva R S, Jesus D R, Pelegrine F (2008). Appl Phys Lett. 93:193115.
[18] Freitas Neto E S, da Silva S W, Morais P C, Vasilevskiy M I, Pereira-da-Silva M A, Dantas N O (2011). J Raman Spectrosc. 42: 1660.
[19] Harrison M T, Kershaw S V, Burt M G, Rogach A L, Kornowski A, Eychmüller A, Weller H (2000). Pure Appl Chem. 72: 295.
[20] Freitas Neto E S, Dantas N O, Barbosa Neto N M, Guedes I, Chen F (2011). Nanotechnology. 22: 105709.
[21] Babentsov V, Riegler J, Schneider J, Ehlert O, Nann T, Fiederle M (2005). J Cryst Growth. 280: 502.
[22] Zhou H, Hofmann D M, Alves H R, Meyer B K (2006). J Appl Phys. 99:103502.
[23] Beaulac R, Archer P I, Van Rijssel J, Meijerink A, Gamelin D R (2008).Nano Lett. 8: 2949.
[24] Lourenço S A, Silva R S, Andrade A A, Dantas N O (2010). J Lumin. 130: 2118.
[25] Lourenço S A, Dias I F L, Poças L C, Duarte J L, de Oliveira J B B, Harmand J C (2003). J Appl Phys. 93: 4475.
[26] Tripathi B, Singh F, Avasthi D K, Das D, Vijay Y K (2007). Physica B (Amsterdam). 400: 70.
[27] Dantas N O, Silva R S, Pelegrini F, Marques G E (2009). Appl Phys Lett. 94: 26310.
[28] Dantas N O, Freitas Neto E S, Silva R S (2010) Diluted Magnetic Semiconductor Nanocrystals in Glass Matrix, Misc: in Nanocrystals (ed. Y. Masuda). Rijeka: InTech. 143 p. Book Chapter
[29] Vossmeyer T, Katsikas L, Giersig M, Popovic I G, Diesner K, Chemsiddine A, Eychmüller A, Weller H (1994). J Phys Chem. 98: 7665.

[30] Jain M K (1991). Diluted Magnetic Semiconductor (ed. M.K.Jain).World Scientific Publishing Co., Singapore. Ch.1, 12 p. Book Chapter.

[31] Cheng Y, Wang Y, Bao F, Chen D (2006). J Phys Chem B. 110: 9448.

[32] Xue H T, Zhao P Q (2009). J Phys D: Appl Phys. 42:015402.

[33] Furdyna J K (1988). J Appl Phys. 64: R29.

[34]Becker W M (1988) Semiconductors and Semimetals, in Diluted Magnetic Semiconductors (ed. J. K. Furdyna and J. Kossut) Academic Press, San Diego,vol. 25; Dietl T (1994), in Handbook on Semiconductors (ed. T. S. Moss and S. Mahajan), North-Holland, Amsterdam.

[35] Lee Y R, Ramdas A K, Aggarwal R L (1988).Phys. Rev. B: Condens. Matter. 38: 10600.

[36] Kim C S, Kim M, Lee S, Kossut J, Furdyna J K, Dobrowolska M (2000). J Cryst Growth. 214–215: 395.

[37] Lee S, Dobrowolska M, Furdyna J K (2005). Phys Rev B: Condens Matter. 72: 075320.

[38] Falk H, Hübner J, Klar P J, Heimbrodt W (2003). Phys. Rev. B: Condens. Matter. 68: 165203; Agekyan V F, Holtz P O, Karczewski G, Moskalenko E S, Yu A (2010). Serov and N. G. Filosofov. Phys Solid State. 52: 27.

[39] Cooley B J, Clark T E, Liu B Z, Eichfeld C M, Dickey E C, Mohney S E, Crooker S A, Samarth N (2009). Nano Lett. 9 : 3142.

[40] Oka Y, Kayanuma K, Shirotori S, Murayama A, Soum I, Chen Z (2002). J Lumin. 100: 175.

[41] Hundt A, Puls J, Henneberger F (2004). Phys Rev B: Condens Matter. 69: 121309(R).

[42] Chernenko A V, Dorozhkin P S, Kulakovskii V D, Brichkin A S, Ivanov S V, Toropov A A (2005). Phys Rev B: Condens Matter. 72: 045302.

[43]Chernenko A V, Brichkin A S, Sobolev N A, Carmo M C (2010). J Phys : Condens Matter. 22: 355306.

[44] Viswanatha R, Pietryga J M, Klimov V I, Crooker S A (2011). Phys Rev Lett. 107: 067402.

[45] Nawrocki M, Rubo Yu G, Lascaray J P, Coquillat D (1995). Phys Rev B: Condens Matter. 52: R2241.

[46] Schmidt T, Scheibner M, Worschech L, Forchel A, Slobodskyy T, Molenkamp L W (2006). J Appl Phys. 100: 123109.

[47] Freitas Neto E S, Dantas N O, Lourenço S A (2012). Phys Chem Chem Phys. 14: 1493.

[48] Dalpian G M, Chelikowsky J R (2006). Phys Rev Lett. 96: 226802.

[49] Du M.-H, Erwin S C, Efros A L, Norris D J (2008). Phys Rev Lett. 100: 179702.

[50] Dalpian G M , Chelikowsky J R (2008). Phys Rev Lett. 100: 179703.

[51] Chan T.-L, Tiago M L, Kaxiras E, Chelikowsky J R (2008). Nano Lett. 8: 596.

[52] Erwin S C (2010). Phys Rev B: Condens Matter. 8: 235433.

[53] Chan T.-L, Kwak H, Eom J.-H, Zhang S B, Chelikowsky J R (2010). Phys Rev B: Condens Matter. 82: 115421.

[54] Chan T.-L, Zayak A T, Dalpian G M, Chelikowsky J R (2009). Phys Rev Lett. 102: 025901.

[55] Rosenthal S J, McBride J, Pennycook S J, Feldman L C (2007).Surf Sci Rep. 62: 111.

# Chapter 2

# Characterization of Nanocrystals Using Spectroscopic Ellipsometry

Peter Petrik

Additional information is available at the end of the chapter

http://dx.doi.org/10.5772/48732

## 1. Introduction

First applications of ellipsometry to the measurement of poly- and nanocrystalline thin films date back to many decades. The most significant step towards the ellipsometric investigation of composite thin films was the realization of the first spectroscopic ellipsometers in the '70s [3, 4, 8], which allowed the measurement of the dielectric function, the imaginary part of which is directly related to the joint density of electronics states sensitively depending upon the changes of the crystal structure. The first models were based on the effective medium approach using constituents of known dielectric functions [5], whereas the volume fraction of the components can be related to the crystal properties of the thin films. This approach is popular ever since, based on its robustness.

The effective medium methods have been followed by a range of different analytical models based on the parameterization of the dielectric function. These models allow the determination of the material properties also in cases when the material cannot be considered as a homogeneous mixture of phases with known dielectric function. These models can also be used for small grains that show size effects (and hence a modified electronic structure and dielectric function), i.e. for grains that can not be modeled by bulk references.

Additional to the nanocrystal properties, the ellipsometric approach allows the sensitive characterization of further layer characteristics like the interface quality (e.g. nanoroughness at the layer boundaries), the lateral or vertical inhomogeneity or the thicknesses in multi-layer structures.

## 2. Basics of ellipsometry

If polarized light will be reflected on the boundary of two media, the state of polarization of the reflected beam will be elliptical, circular, or linear depending on the properties of the sample. In most cases, the reflected light is elliptically polarized, that's why the method is called ellipsometry. Ellipsometry directly measures the change of polarization caused by the reflection, i.e. the complex reflectance ratio defined by

$$\rho = \frac{\chi_r}{\chi_i} = \frac{r_p}{r_s} = \tan\Psi e^{i\Delta}, \qquad (1)$$

where $\chi_r = E_{r,p}/E_{r,s}$ and $\chi_i = E_{i,p}/E_{i,s}$ ($r$: reflected, $i$: incident; $p$: parallel to the plane of incidence; $s$: perpendicular to the plane of incidence) are the states of polarization with $E = E_0 e^{i(\omega t+\delta)} e^{-i\omega \frac{N}{c} x}$ describing the electromagnetic plane wave ($\omega$: angular frequency; t: time; $\delta$: phase; $N$: complex refractive index; $c$: speed of light; $x$: position), $r_p$ and $r_s$ are the reflection coefficients (the ratio of reflected and incident $E$ values) of light polarized parallel and perpendicular to the plane of incidence, respectively, while $\Psi$ and $\Delta$ are the ellipsometric angles [15, 17]. The latter are the raw data of an ellipsometric measurement usually plotted as a function of wavelength, angle of incidence, or time.

The sample properties like the layer thickness or the refractive index are determined using optical models, in which one assumes a layer structure and the refractive index of each layer at each wavelength. Using such optical models, the $\Psi$ and $\Delta$ values can be calculated (for each wavelength and angle of incidence) and compared with the measured $\Psi$ and $\Delta$ spectra. Next, the parameters of the optical model (e.g. thickness, refractive index, or parameters of the dispersion) are fitted using optimization algorithms like the Levenberg-Marquardt method in order to minimize the difference between the measured and calculated $\Psi$ and $\Delta$ values.

## 3. Measurable nanocrystal properties

Ellipsometry can measure nanocrystals in a thin film form. The optical model consists of a substrate with usually known optical properties (e.g. single-crystalline silicon or glass) and one or more thin films containing the nanocrystals. The information needed to calculate the ellipsometric angles $\Psi$ and $\Delta$ (that in turn can be compared with the measured angles) are the angle of incidence, the thickness(es) of the layer(s) and the complex refractive indices of each medium. The square of the refractive index is the dielectric function, the imaginary part ($\epsilon_2$) of which is directly proportional to the joint density of electronic states of the measured crystalline or partly crystalline material. Compared to the typical sensitivities ($< 10^{-3}$ in $\epsilon$) the difference between the dielectric functions of the amorphous and single-crystalline phases are huge, as shown in Fig. 1. It is clearly seen that the dielectric functions are largely different depending on the crystallinity of silicon ranging from single-crystalline through nanocrystalline to amorphous. Note that the sensitivity is especially high around the critical point energies, which appear as peaks in $\epsilon_2$. For example, the difference of the three cases in $\epsilon_2$ is larger than 10 at the critical point energy of 4.2 eV.

The key of measuring nanocrystals is to relate the dielectric function to nanocrystal properties. The technique that is most widely used for more than 30 years is the effective medium theory [5, 10]. In this theory it is assumed that the material is a mixture of phases large enough to retain their bulk-like properties, but smaller than the wavelength of the probing light, so that scattering can be avoided. In case of a mixture of two components with dielectric functions of $\epsilon_a$ and $\epsilon_b$ the effective dielectric function ($\epsilon$) can be determined using effective medium theory, the most general form of which is the self-consistent Bruggeman effective medium

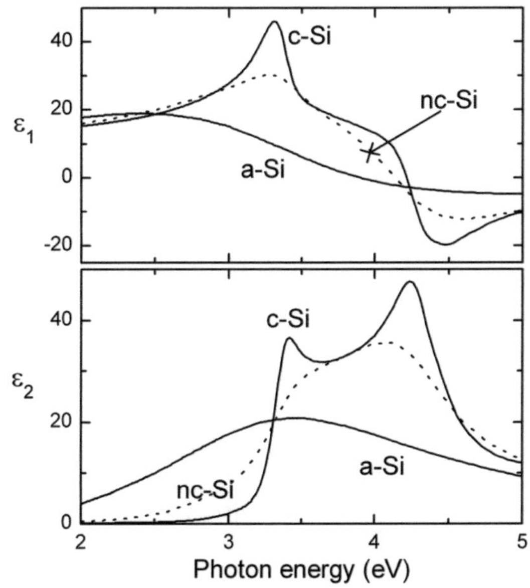


**Figure 1.** Real ($\epsilon_1$) and imaginary ($\epsilon_2$ parts of the dielectric function of single-crystalline (c-Si), amorphous (a-Si), and nanocrystalline (nc-Si) silicon.

approximation (B-EMA):

$$0 = f_a \frac{\epsilon_a - \epsilon}{\epsilon_a + 2\epsilon} + f_b \frac{\epsilon_b - \epsilon}{\epsilon_b + 2\epsilon}. \quad (2)$$

The fit parameters of such models are the volume fractions $f_a$ and $f_b$ of components $a$ and $b$, respectively ($\epsilon_a$ and $\epsilon_b$ are known, $\epsilon$ is calculated assuming $f_a$ and $f_b$ values). In turn, the change of the volume fractions can be related to the morphological changes of the material.

As an example of using B-EMA, lets consider an intensively investigated nanocrystalline material, namely the nanocrystalline silicon prepared by a range of deposition techniques from low temperature chemical vapor deposition to magnetron sputtering. A typical optical model for the measurement of this material uses components of single-crystalline silicon (c-Si, dielectric function from Ref. [16]), amorphous silicon (a-Si, dielectric function from Ref. [6]), nanocrystalline silicon (nc-Si, dielectric function from Ref. [18]), and voids. Some optical models for large-grain (Model 1) and small-grain (Model 2) nanocrystalline silicon layers deposited on oxidized single-crystalline silicon wafers are shown in Fig. 2 (see also Ref. [25]). The surface nanoroughness is also considered as a homogeneous layer [9], the thickness of which correlates with the root mean square roughness [22]. There are more sophisticated theories to analyze the surface roughness from the polarized optical response [14], but the robustness of the above effective medium approach makes it a popular, most widely used method. Note however, that the accurate, validated determination of the root mean square roughness is a problem even by scanning probe methods, because the obtained numerical value may strongly depend on the measurement configuration (e.g. the window size or the quality of the tip, [22]).

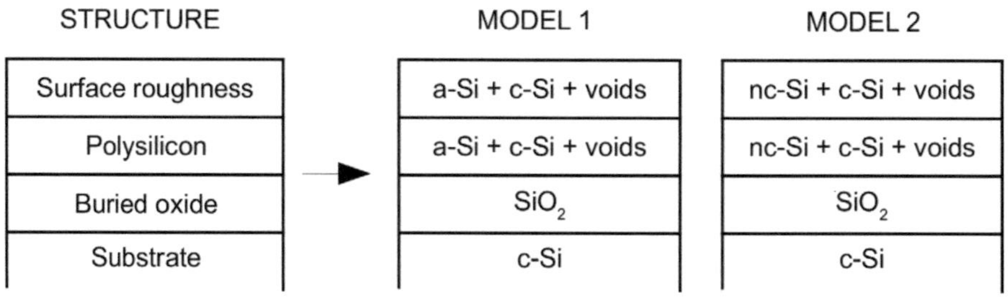


**Figure 2.** Some B-EMA models for nanocrystalline Si on c-Si with different expected grain sizes. Models 1 and 2 can be used for large and small grain sizes, respectively.

In Fig. 2 both the surface nanoroughness and the bulk layer properties are described by a combination of a-Si, c-Si, and voids or nc-Si, c-Si, and voids. The voids ratio serves as a density correction in the bulk layer, and it describes the character of the surface nanoroughness in the roughness layer. In many cases, the void ratio is fixed at 50 % due to a possible correlation with the thickness of the surface roughness layer. The model also contains the buried thermal oxide. The thickness of this oxide layer can also be determined in one fitting step together with the volume fractions of the components in the roughness and bulk nanocrystalline silicon layer, as well as their layer thicknesses.

For layers with smaller nanocrystals (several times 10 nanometers or below) the use of the nc-Si reference dielectric function [18] provides a better fit than using the conventional components of c-Si, a-Si, and voids. The volume fraction of the nc-Si component in the model is an indication of the crystallinity (Fig. 3). The higher the volume fraction of the nc-Si component, the smaller the grain size [25]. The sensitivity is revealed to be high, however, quantitative measurements verified by reference methods were not performed. Micrographs of transmission electron microscopy were available, but the technique of a reliable grain size analysis using image processing is missing.

There are a range of methods for the analytical representation of the dielectric function of semiconductors. A general empirical approach is to use a combination of Lorentz oscillators. This method allows the determination of the critical point energies and the layer thickness [13].

The critical point features can also be characterized by calculating the derivative of the dielectric function [7]. Using this approach the broadening of the critical point features can especially well be characterized. In turn, the grain size can indirectly be measured, because the broadening is proportional to the electron scattering at the grain boundaries, i.e. the amount of grain boundaries is measured.

The generalized oscillator model applies a standard analytical equation for all types of critical points [11]. The dimensionality of the critical point can be adjusted by a parameter of the model.

Another general approach for the parameterization of the critical point features of the dielectric function is the generalized critical point model suggested by B. Johs et al. [19]. In this model the dielectric function around the critical point is described by four Gaussian-broadened polynomials. A detailed description can be found in Ref. [12].

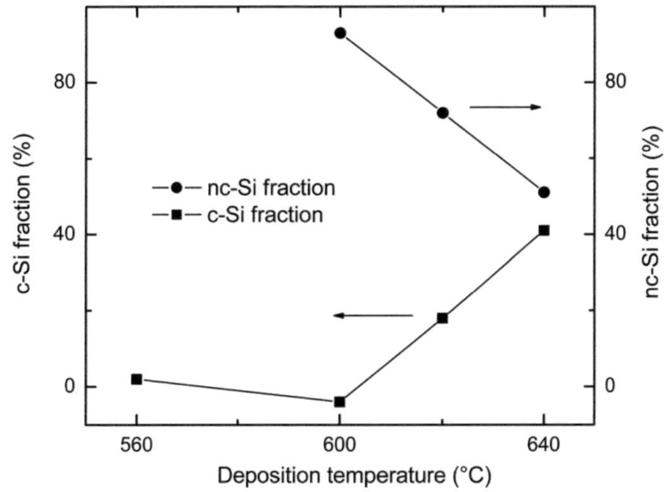


**Figure 3.** c-Si and nc-Si volume fractions as functions of the deposition temperature. The grain size increases with increasing temperature. Reprinted with permission from Journal of Applied Physics 87, P. Petrik, Ellipsometric study of polycrystalline silicon films prepared by low-pressure chemical vapor deposition, 1734 (2000). Copyright 2000, American Institute of Physics.

A special parameterization has been developed by S. Adachi. This model dielectric function describes each critical point with a specific analytical equation depending on the type of the critical point [1, 26]. As a result, one has a set of Kramers-Kroenig-consistent analytical equations, with each equation describing an oscillator. The superposition of all the fitted oscillators results in the final dielectric function as shown in Fig. 4. It is also shown that the effective medium approximation using the combination of c-Si and a-Si as described above doesn't allow an acceptable fit (short-dashed line). Although the number of parameters to fit is large, there are a lot of possibilities to couple or fix non-sensitive parameters [23].

The most important properties of the nanocrystal layers that can be measured by ellipsometry are the layer thickness and the grain size (Table 1). The crystallinity can be defined for ellipsometry as the "position" of the dielectric function between that of the single-crystalline and the amorphous ones. In terms of effective medium theory this can be defined as the c-Si to a-Si ratio of the model components. The density is also a parameter that can be expressed compared to a reference value. Staying at the above example the model "c-Si + a-Si + void" compares the density to the purely "c-Si + a-Si" case.

Additional to the above parameters a range of thin film-related parameters can be determined. Most important may be the vertical inhomogeneity and the interface quality. Using ellipsometry it can sensitively be checked whether the layer is of optical quality.

## 4. Applications

The range of ellipsometric applications for nanocrystals is very large. The most important applications comprise semiconductor nanocrystals, with silicon still being the number one.

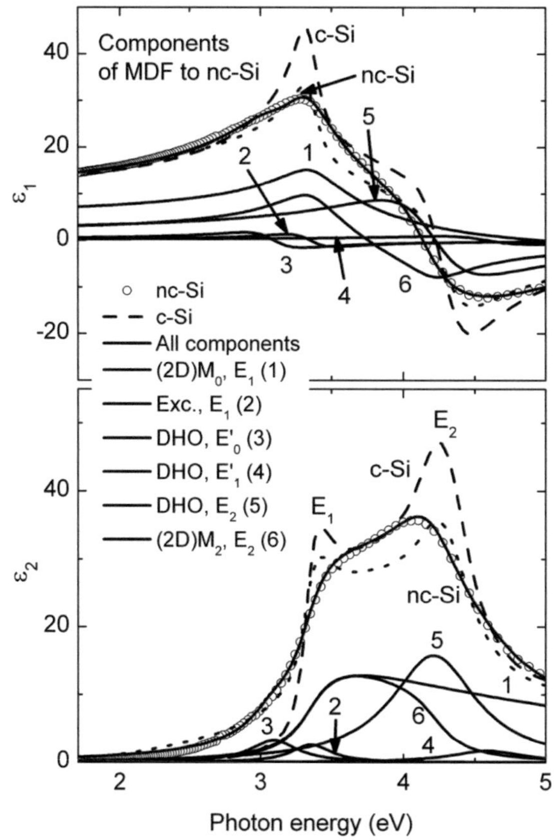


**Figure 4.** Oscillators of the model dielectric function of S. Adachi (numbered solid lines) fitted to the reference dielectric function of a fine-grained polycrystalline silicon (nc-Si). The long-dashed line is the single-crystalline silicon (c-Si) reference. The short-dashed line shows the fit using the effective medium approximation described above with the components of c-Si and a-Si.

In this section some relevant applications known to the author are selected, with no claim to completeness.

The correlation between the grain size and the broadening of the dielectric function at the $E_1$ critical point was revealed by H. V. Nguyen and R. W. Collins in an in situ study during plasma enhanced chemical vapor deposition [21]. The grain size was determined from the layer thickness assuming a three-dimensional, isolated particle growth. It was shown that the theoretically predicted relationship that the broadening is proportional to $1/d$, where $d$ denotes the grain size, holds for the investigated nanocrystals (Fig. 5). Furthermore, the extrapolation of $d \rightarrow \infty$ results in the expected broadening value of the single-crystalline silicon, which suggests that this behavior is valid for smaller crystals as well. It is important to note that the investigated thickness range of 20-25 nm was large enough to neglect the shift of the critical point energy due to the quantum confinement.

G. F. Feng and R. Zallen [13] used Raman spectroscopy to verify the linearity between broadening and reciprocal grain size (Fig. 6) on nanocrystals created by ion implantation in

| Property | Notes |
|---|---|
| Grain size | Sub-nanometer sensitivity<br>Indirect: "calibration" needed |
| Crystallinity | Amount of grain boundaries<br>In case of one-phase layers<br>indicative to grain size |
| Layer thickness | Sub-nanometer sensitivity<br>Also in multi-layers |
| Density | Compared to the reference materials<br>used in the model |
| Inhomogeneity | Both vertical and lateral |
| Interface quality | Whether one needs an additional<br>transition layer to achieve a good fit |
| Surface nanoroughness | As a homogeneous roughness layer<br>or using more sophisticated methods |
| Shape of crystals | In principle possible<br>using anisotropy and B-EMA screening |

**Table 1.** Properties of nanocrystal layers that can be measured by ellipsometry.

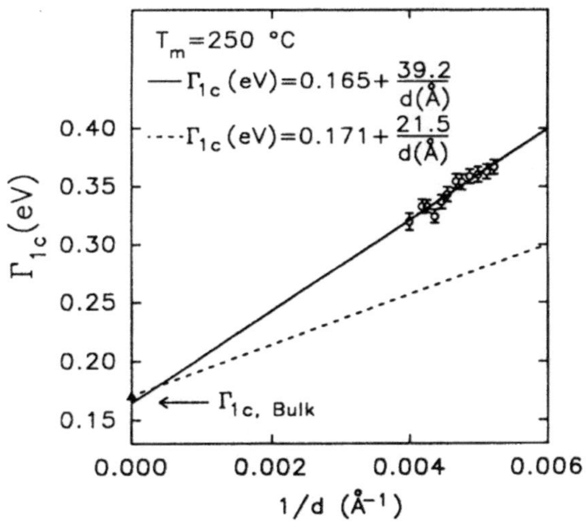


**Figure 5.** Broadening of the dielectric function at the $E_1$ critical point energy for grain sizes of $d$ measured during layer growth. Reprinted from Physical Review B 47, H. V. Nguyen and R. V. Collins, Finite-size effects on the optical functions of silicon microcrystallites: A real-time spectroscopic ellipsometry study, 1911 (1993). Copyright 1993, American Physical Society.

GaAs in the size range of 5-50 nm. A good linearity was found for all investigated critical point energies. This example also proves that the sensitivity is high enough, and the quantitative determination of the grain size is possible by validation using a reference method.

A further example of a nanocrystalline structure is porous silicon prepared by electrochemical etching. After etching, the sample has a well-defined nanostructure depending on the etching conditions. The size of the remaining nanocrystals strongly depends on the doping level of the substrate. In the study of P. Petrik et al. [23] substrates with different B-doping were used

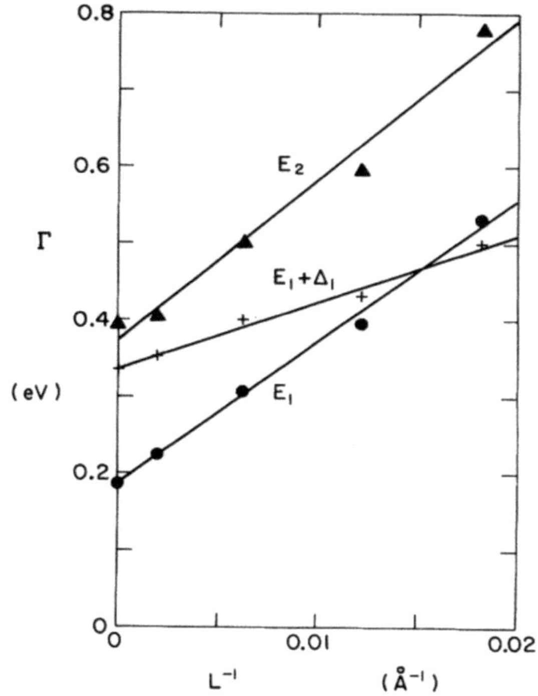


**Figure 6.** Broadening of the dielectric function at different critical points as a function of the reciprocal crystallite size in ion implanted GaAs. Reprinted from Physical Review B 40, G. F. Feng and R. Zallen, Optical properties of ion-impanted GaAs: The observation of finite-size effects in GaAs microcrystals, 1064 (1989). Copyright 1989, American Physical Society.B

to achieve substrate resistivities from 0.01 to 0.09 Ωcm and nanocrystal sizes from 3 to 12 nm. In case of porous silicon the use of proper optical models is of crucial importance, because the layers are usually vertically non-uniform. In this study the dielectric function of the porous silicon layers were analyzed using the model dielectric function of S. Adachi described above. The fitted broadening parameters of the model are plotted as a function of the grain size in Fig. 7. Similar to the above cases, a more or less systematic behavior was found also in this study, especially for the $E_1$ critical point ($\Gamma_1$). The reason is that the porous silicon layers are vertically inhomogeneous, and the information depth (in c-Si) at the $E_1$ and $E_2$ critical point energies are about 30 and 10 nm, respectively. So $\Gamma_2$ is very sensitive to surface deteriorations. The dependence on the layer thickness is also due to the vertical gradient in the nanostructure.

Chemical vapor deposition combined with oxidation was used to prepare nanocrystal layers with well-defined grain sizes [20]. The thin nanocrystal films were deposited on quarz substrates, and the nanocrystal size was set by the layer thickness and the time of the oxidation process, with nanocrystal sizes verified by electron microscopy and X-ray diffraction. For the analysis of the nanocrystal layers the above mentioned effective medium approximation could successfully be applied using a combination of nc-Si, c-Si, and $SiO_2$ reference dielectric functions [2]. It was shown that the nanocrystallinity defined by $f_{nc-Si}/(f_{c-Si} + f_{nc-Si})$, where $f$ denotes the fitted volume fractions of the components (see equation 2), correlates well with the nanocrystal size determined by the layer thickness (see Fig. 8). This example

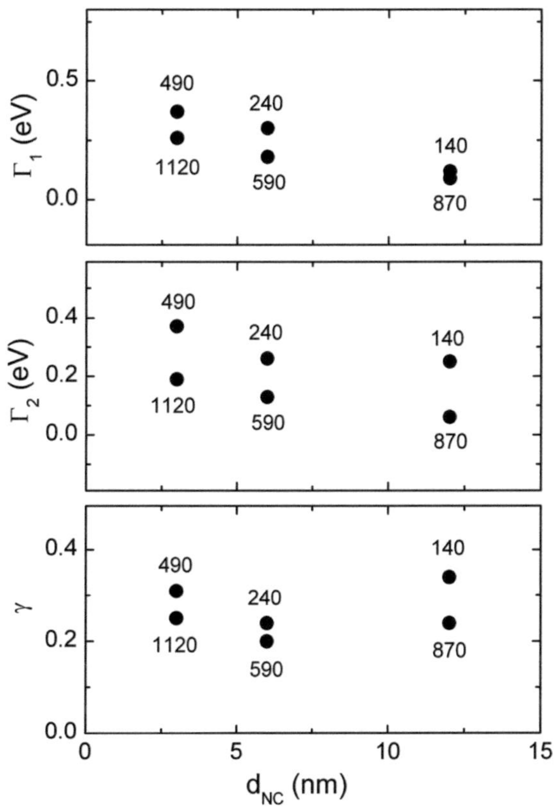


**Figure 7.** Broadening values measured using the model dielectric function of S. Adachi for different grain sizes ($d_{NC}$). The numbers at the measured points show the thicknesses of the measured layers. Reprinted from Journal of Applied Physics 105, P. Petrik et al., Nanocrystal characterization by ellipsometry in porous silicon using model dielectric function, 024908 (2009). Copyright 2009, American Institute of Physics.

also shows that the nanocrystal size can sensitively (though indirectly) be measured by this optical technique.

The importance of an optical technique is not only the speed and sensitivity, but also the in situ capability. Using special beam-guiding techniques the ellipsometric measurement can even be made in a vertical furnace through the base plate (i.e. without modifying the furnace walls) [24]. The performance of current computers allows a real time evaluation of the process, providing the layer thicknesses and the information on the morphology. Fig. 9 shows the results of an in situ measurement made in a vertical furnace during crystallization of a deposited amorphous layer. The thickness of the amorphous silicon layer was about 40 nm, deposited on a thermally oxidized (10 nm oxide) silicon wafer. Accordingly, the optical model consists of a single-crystalline silicon substrate, a buried $SiO_2$ layer (with a thermal $SiO_2$ reference dielectric function), an initially amorphous silicon layer described by the a-Si reference, and an $SiO_2$ surface oxide layer. The surface roughness could be neglected for the deposited amorphous silicon layer. During annealing crystals are formed in the amorphous silicon layer. This can be followed by a simple effective medium combination of components

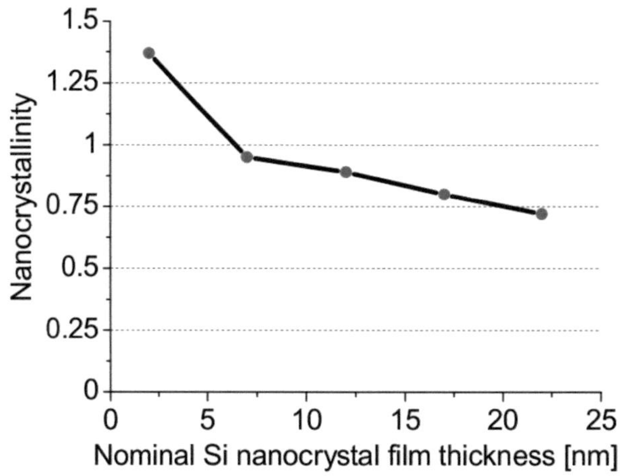


**Figure 8.** Nanocrystallinity as a function of the thickness of nanocrystal films prepared by chemical vapor deposition and oxidation of quartz substrates. Reprinted from Mater. Res. Soc. Symp. Proc. 1321, E. Agocs et al., Optical characterization using ellipsometry of si nanocrystal thin layers embedded in silicon oxide, DOI: 10.1557/opl.2011.949. Copyright 2011, Cambridge University Press.

a-Si and c-Si, as described above. The increasing volume fraction of c-Si represents the process of crystallization during annealing (see Fig. 9).

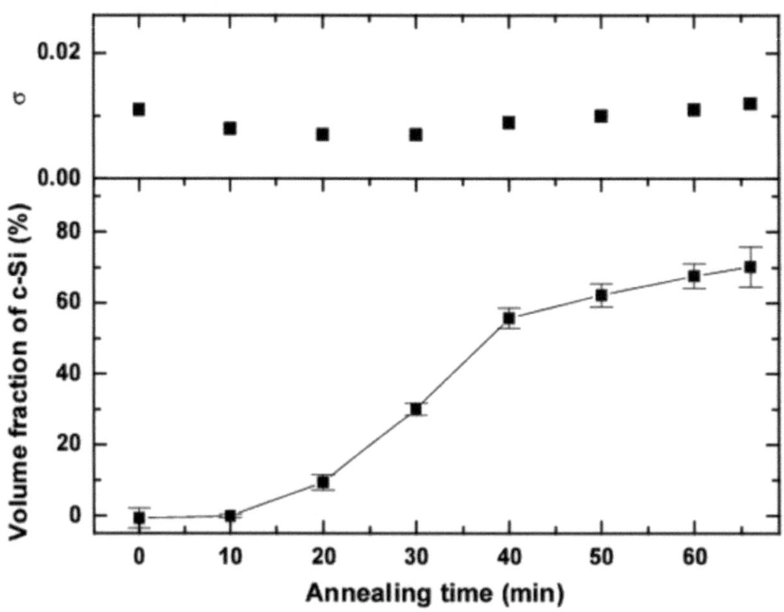


**Figure 9.** Fitted volume fraction of crystalline silicon in the effective medium model as a function of annealing time. Reprinted from Thin Solid Films 364, P. Petrik et al, In situ spectroscopic ellipsometry for the characterization of polysilicon formation inside a vertical furnace, 150 (2000). Copyright 2000, with permission from Elsevier.

## 5. Conclusions

The capability of measuring the change of phase by the reflection of polarized light on a surface or layer structure allows a typical sensitivity of less than one nanometer for the layer thickness and $10^{-3} - 10^{-4}$ for the refractive index. These capabilities together with a spectroscopic use in the photon energy range of the characteristic interband transitions makes the method applicable for an accurate determination of nanocrystal properties especially for Si, but also for wide band gap semiconductors using an extension into the deep UV spectral range. The sensitivity is high down to the smallest nanocrystal sizes. The requirement for an accurate nanocrystal measurement is a layered form with optical quality interfaces. The thickness, homogeneity and interface qualities of the layers can be measured directly, whereas the properties related to the nanocrystal structure (like the crystallinity, the nanocrystal size or the density of the layer) can be obtained indirectly using proper optical models and "calibration" by reference methods.

## Author details

Peter Petrik
*Institute for Technical Physics and Materials Science (MFA), Research Centre for Natural Sciences, Konkoly Thege u. 29-33, 1121 Budapest, Hungary*

## 6. References

[1] Adachi, S. [1988]. Model dielectric constants of si and ge, *Phys. Rev. B* 38: 12966.
[2] Agocs, E., Petrik, P., Fried, M. & Nassiopoulou, A. G. [2011]. Optical characterization using ellipsometry of si nanocrystal thin layers embedded in silicon oxide, *Mater. Res. Soc. Symp. Proc.* 1321: DOI: 10.1557/opl.2011.949.
[3] Aspnes, D. E. [1975]. Extended spectroscopy with high-resolution scanning ellipsometry, *Physical Review B* 12: 4008.
[4] Aspnes, D. E. [1981]. Studies of surface, thin film and interface properties by automatic spectroscopic ellipsometry, *SPIE Proc.* 276: 312.
[5] Aspnes, D. E. [1982]. Optical properties of thin films, *Thin Solid Films* 89: 249.
[6] Aspnes, D. E. [1985]. The accurate determination of optical properties by ellipsometry, *in* E. D. Palik (ed.), *Handbook of Optical Constants of Solids*, Academic Press, New York.
[7] Aspnes, D. E., Kelso, S., Olson, C. & Lynch, D. [1982]. *Phys. Rev. Lett.* 48: 1863.
[8] Aspnes, D. E. & Studna, A. A. [1978]. Methods for drift stabilization and photomiltiplier linearization for photometric ellipsometers and polarimeters, *Review of Scientific Instruments* 49: 291.
[9] Aspnes, D. E., Theeten, J. B. & Hottier, F. [1979]. Investigation of effective-medium models of microscopic surface roughness by spectroscopic ellipsometry, *Phys. Rev. B* 20: 3292.
[10] Bruggeman, D. A. G. [1935]. Berechnung verschiedener physikalischer konstanten von heterogenen substanzen i. dielektrizitätskonstanten und leitfähigkeiten der mischkörper aus isotropen substanzen, *Ann. Phys. (Lepzig)* 24: 636.
[11] Cardona, M. [1969]. Modulation spectroscopy, *in* F. Seitz, D. Turnbull & H. Ehrenreich (eds), *Solid State Physics*, Academic Press, New York.
[12] Collins, R. W. & Ferlauto, A. S. [2005]. Optical physics of materials, *in* E. G. Irene & H. G. Tomkins (eds), *Handbook of ellipsometry*, William Andrew, Norwich, NY.

[13] Feng, G. F. & Zallen, R. [1989]. Optical properties of ion-implanted gaas: The observation of finite-size effects in gaas microcrystals, *Physical Review B* 40: 1064.

[14] Franta, D., Ohlidal, I. & Necas, D. [2008]. Optical quantities of rough films calculated by rayleigh-rice theory, *Physica Status Solidi C* 5: 1395.

[15] Fujiwara, H. [2007]. *Spectroscopic Ellipsometry*, Wiley, Chichester.

[16] Herzinger, C. M., Johs, B., McGahan, W. A., Woollam, J. A. & Paulson, W. [1998]. Ellipsometric determination of optical constants for silicon and thermally grown silicon dioxide via a multi-sample, multi-wavelengt, multi-angle investigation, *J. Appl. Phys.* 83: 3323.

[17] Irene, E. A. & Tompkins, H. G. [2005]. $Sio_2$ films, *in* E. G. Irene & H. G. Tomkins (eds), *Handbook of ellipsometry*, William Andrew, Norwich, NY.

[18] Jellison, G. E., Jr., Chisholm, M. F. & Gorbatkin, S. M. [1993]. Optical functions of chemical vapor deposited thin-film silicon determined by spectroscopic ellipsometry, *Appl. Phys. Lett.* 62: 348.

[19] Johs, B., Herzinger, C. M., Dinan, J. H., Cornfeld, A. & Benson, J. D. [1998]. *Thin Solid Films* 313: 137.

[20] Lioutas, C. B., Vouroutzus, N., Tsaioussis, I., Frangis, N., Gardelis, S. & Nassiopoulou, A. G. [2008]. Columnar growth of ultra-thin nanocrystalline si films on quartz by low pressure chemical vapor deposition: accurate control of vertical size, *Physica Status Solidi a* 11: 2615.

[21] Nguyen, H. V. & Collins, R. W. [1993]. Finite-size effects on the optical functions of silicon microcrystallites: A real-time spectroscopic ellipsometry study, *Physical Review B* 47: 1911.

[22] Petrik, P., Biró, L. P., Fried, M., Lohner, T., Berger, R., Schneider, C., Gyulai, J. & Ryssel, H. [1998]. Surface roughness measurement on polysilicon produced by low pressure chemical vapor deposition using spectroscopic ellipsometry and atomic force microscopy, *Thin Solid Films* 315: 186.

[23] Petrik, P., Fried, M., Vazsonyi, E., Basa, P., Lohner, T., Kozma, P. & Makkai, Z. [2009]. Nanocrystal characterization by ellipsometry in porous silicon using model dielectric function, *J. Appl. Phys.* 105: 024908.

[24] Petrik, P., Lehnert, W., Schneider, C., Fried, M., Lohner, T., Gyulai, J. & Ryssel, H. [2000]. In situ spectroscopic ellipsometry for the characterization of polysilicon formation inside a vertical furnace, *Thin Solid Films* 364: 150.

[25] Petrik, P., Lohner, T., Fried, M., Biro, L. P., Khanh, N. Q., Gyulai, J., Lehnert, W., Schneider, C. & Ryssel, H. [2000]. Ellipsometric study of polycrystalline silicon films prepared by low pressure chemical vapor deposition, *J. Appl. Phys.* 87: 1734.

[26] Tsunoda, K., Adachi, S. & Takahashi, M. [2002]. Spectroscopic ellipsometry study of ion-implanted si(100) wafers, *J. Appl. Phys.* 91: 2936.

# Chapter 3

# Localized Nano-Environment for Integration and Optimum Functionalization of Chlorophyll-*a* Molecules

P. Vengadesh

Additional information is available at the end of the chapter

http://dx.doi.org/10.5772/48707

## 1. Introduction

In recent years, there has been a growing trend of applying thin films containing natural pigments and dyes to various devices [1, 2] such as photovoltaic cells, optical waveguides and ultrafast optical switches in basic research, aiming at high performance and quantum efficiency. Though still at an experimental stage, much interest has been directed towards incorporating such photoactive materials into functionalized synthetic mediums or "artificial membranes" (AM). The photosynthetic biomaterials (PBMs) of interest such as the abundant Chlorophyll (Chl), in its native environment are contained within thylakoid membranes meant to provide a suitable host medium for optimum photoactivity to take place. Through millions of years of evolution, these membranes are constantly "rebuilt" and optimized to allow localized nano-environments for the integration of the synthesized PBMs. These nano-environments effectively introduce a higher area of light and nutrient absorption for Chl functionality. Inability to artificially replicate such localized nano-environment within the AM for nano-encapsulation of Chl molecules may be the key element hindering further progress in research into reconstituted PBMs for development of photovoltaic based devices and sensors.

Generally, suitable AM for impregnation of PBMs intended for photovoltaic studies require high levels of visible light transparency and penetration for photoreaction to occur efficiently and effectively. It is also of concern of making sure that the AM is totally inert to the active material used besides providing physical integrity, cellular-like viscosity, elasticity and humidity. A very important feature, internal humidity is required for optimum operation of most PBMs. For such a reason, AMs are required to have the ability to lock-in humidity and maintain a "gel-like" property similar to the natural lipid based membrane. By creating an almost natural-like environment, photosensitive biomaterials such as the Chl molecules were shown to be able to retain their intrinsic properties for extended periods of time [3]. This is especially important due to the fact that Chl molecules are composed of extremely unstable compounds and therefore are readily destroyed by continuous exposure to light, oxygen, heat and acidic and alkaline substances. While most of the works [4] carried-out are concentrated into replicating photoswitching effects as in natural environment, not much effort was given into providing an in-depth comparative discussion into the effects of encapsulation of Chl within the AM matrix and photoresponse. The crucial the requirement is to create a compatible host medium for optimum photoreaction [3]. Such synthetic host environments are required to emulate the conventional tasks of natural lipid-based membranes. Generally, this involves electrons (and hydrogen ions) involved in the photosynthesis chain and nutrients that are transported based on two major stimulating responses, a trans-membrane concentration gradient [5] and upon light activation of photosensitive compounds [6].

In order to facilitate such complex processes, the material chosen for development of the synthetic host medium must recreate and maintain internal cellular viscosity. This would allow similar photo-induced steps to be maintained efficiently within the physical structure of the synthetic cell provided any loss of internal humidity is minimized or regulated by means of timed moisture absorption. In nature, the physical and viscous integrity of the plant cell is maintained by regulated fluid intake and elimination. Many different polymer materials offer the ability to be re-engineered to modify its' physical properties to mimic the native properties. One such polymer is carboxymethyl cellulose (CMC), which can be prepared in the form of an aqueous "gel-like" viscous solution with characteristics similar to a biological cell. CMC has been used in various practical fields [7, 8, 9]. In one such application [10], He Huang *et al.* studied electrochemical properties of proteins or enzymes such as Heme protein at CMC modified electrodes. They reported that CMC is able to provide a new and different matrix for immobilization of proteins.

## 2. Carboxymethyl cellulose as artificial membrane

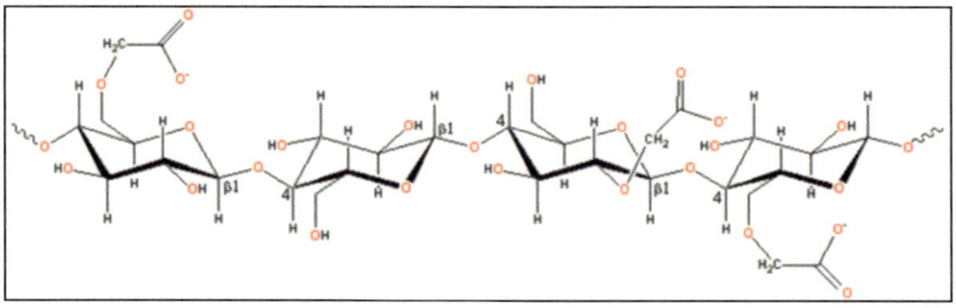

**Figure 1.** Structural unit of CMC [12].

Cellulose, the structural material of plant cell walls [11] is composed of repeating D-glucose units linked through ß-1, 4 glycosidic bonds (as shown in Figure 1). Under normal conditions, a tight packing of polymer chains is shown resulting in a highly crystalline structure. Such crystalline structures are able to resist salvation in aqueous media as a result of existence of hydroxymethyl groups of anhydroglucose residues found above and below the plane of the polymer backbone. In order to increase water solubility, cellulose is treated with alkali resulting in swelling of the structure. This intermediate structure would then be reacted with chloroacetic acid, methyl chloride or propylene oxide to produce CMC, methyl cellulose (MC), hydroxypropyl cellulose (HPMC) or hydroxypropyl cellulose (HPC).

Recently, CMC have been extensively investigated for its potential as an AM material due to its desirable properties such as nontoxicity, biocompatibility, high hydrophilicity and excellent film forming abilities besides being not too expensive for large-scale commercial usage [13, 14]. Other positive factors are that these materials are non-ionic and compatible with surfactants, other water-soluble polysaccharides and salt [11]. CMC and its derivatives can also be dissolved in aqueous or aqueous-ethanol solutions for specific requirements. The resulting materials produced are generally odorless and tasteless, flexible and are of moderate strength, moderate to moisture and oxygen transmission, resistant to fats and oils [15-17]. CMC exhibits good film forming and suitable host medium properties. It is an excellent gas barrier material while at the same time allows water vapor diffusion. Many CMC based derivates have water vapor permeability comparable to Low Density Polyethylene (LDPE). Thus, their permeability or mechanical properties can be fine-tuned as dictated by the need of a specific application.

Proper understanding of various photochemical and photophysical properties of photosynthetic pigments in organic polymer matrix as an "in-vitro" biomimetic immobilizing media or AM, such as CMC may enable development of prototypes of PBM based devices. Advantages are attributed to more rigid stability and ability to control dispersion of incorporated photosynthetic molecules. CMC also acts as an effective dispersion agent for the active material and had been extensively studied and reviewed [18-23]. Formed as a film on top of the photosynthetic pigments, CMC are also observed to preserve photosensitivity for extended periods of time [3] while allowing crucial transfer of oxygen and moisture. Its potential functions and applications to film formation, coating and as AM can be achieved upon further extensive research on the methods of film and matrix formation and improvement of its internal properties. This would be crucial to warrant the increased need of considerations in potential functions and applications of CMC, especially involving Chl-*a* molecules when incorporated into the CMC matrix.

## 3. Photosynthetic biomaterial: Chlorophyll-*a*

Photosynthetic chloroplasts containing Chl molecules are found in green tissues of higher plants. They are mostly found in the form of Chl-*a* ($C_{55}H_{72}MgN_4O_5$) and Chl-*b* ($C_{55}H_{70}MgN_4O_6$), while other forms in nature are to be found as Chl-*c* ($C_{35}H_{30}MgN_4O_5$) and Chl-*d* ($C_{54}H_{70}MgN_4O_6$). These structures (Figure 2) differ only in the substituent at C-7, which is $-CH_3$ in Chl-*a* but –CHO in Chl-*b*. For Chl-*d* meanwhile, it is propionic acid substituent at C-17, which is isoprenoid alcohol phytol achieved by means of the esterification process.

Improvements over solid-state diagnostic methods in recent history enabled photosynthesis to be extensively studied in significant detail. Of particular interest are the photochemical and photophysical properties of photosynthetic pigments. Photosynthetic systems containing Chl-*a* molecules are popularly used for development of novel designs implementing the light-mediated processes rather than just for solar energy conversion [25]. Chl-*a* is considered significant for pigment studies in connection with the primary photosynthetic process, as well as light energy conversion. Chl exhibits a functional duality as energy collector and primary electron carrier since it plays an important role in the photosynthesis process occurring in plants and in some bacteria. Therefore, the absorption spectra of green plants have shown multiple forms of Chl. The molecules are arranged in a highly ordered state on grana thylakoid membranes and their local concentration of porphyrin rings is relatively high.

| | Chlorophyll *a* | Chlorophyll *b* | Chlorophyll *c1* | Chlorophyll *c2* | Chlorophyll *d* |
|---|---|---|---|---|---|
| Molecular formula | $C_{55}H_{72}O_5N_4Mg$ | $C_{55}H_{70}O_6N_4Mg$ | $C_{35}H_{30}O_5N_4Mg$ | $C_{35}H_{28}O_5N_4Mg$ | $C_{54}H_{70}O_6N_4Mg$ |
| C3 group | $-CH=CH_2$ | $-CH=CH_2$ | $-CH=CH_2$ | $-CH=CH_2$ | -CHO |
| C7 group | $-CH_3$ | -CHO | $-CH_3$ | $-CH_3$ | $-CH_3$ |
| C8 group | $-CH_2CH_3$ | $-CH_2CH_3$ | $-CH_2CH_3$ | $-CH=CH_2$ | $-CH_2CH_3$ |
| C17 group | $-CH_2CH_2COO-$ Phytyl | $-CH_2CH_2COO-$ Phytyl | -CH=CHCOOH | -CH=CHCOOH | $-CH_2CH_2COO-$ Phytyl |
| C17-C18 bond | Single | Single | Double | Double | Single |
| Occurrence | Universal | Mostly plants | Various algae | Various algae | cyanobacteria |

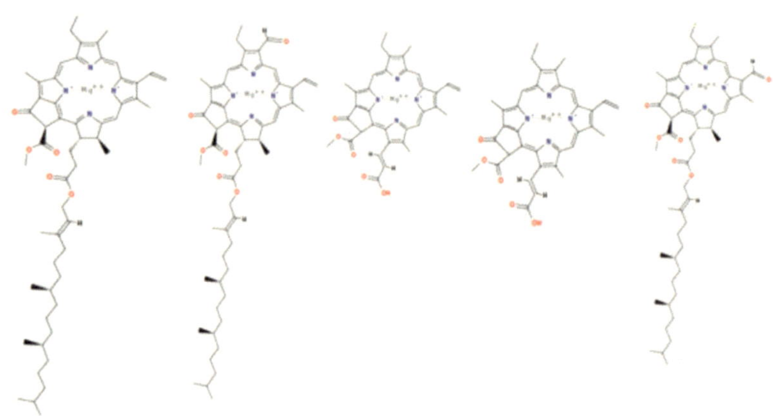

**Figure 2.** Structure of chlorophylls [24].

Photosynthetic organisms are able to oxidize water, upon absorption of sunlight, to produce oxygen. In higher plants this process takes place in the thylakoid membranes located inside the chloroplast. These membranes have two key reaction centers, P700 and P680, both of which operate in two distinct photosystems, photosystem I (PSI) and II (PSII) and consist of Chl-*a* molecules. The oligomeric nature of P680 *"in-vivo"* is a matter of considerable interest. Some studies suggest that P680 composed of one molecule of Chl-*a* [26-28], while others proposed it as a dimer [29-31]. Absorption and fluorescence spectra of dry Langmuir-Blodgett multilayers

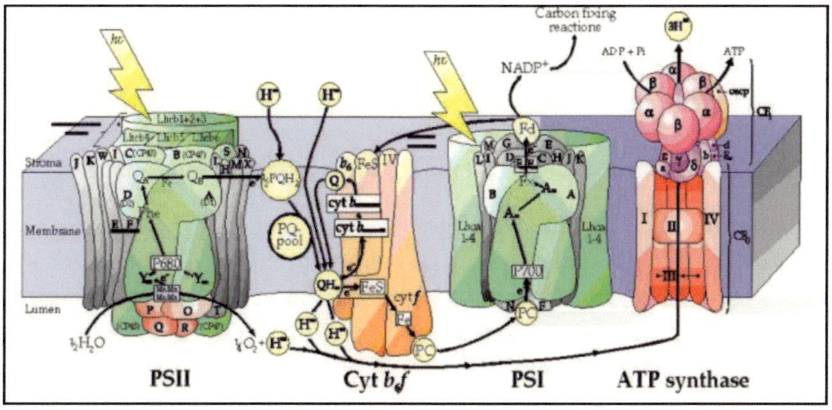


**Figure 3.** Simple illustration showing the PSI and PSII processes involved in photosynthesis [32].

of Chl-*a* present similarities with those observed for the P680 or P700 and its electron acceptors make up PSI, whereas P680 and its electron carriers make up PSII as shown in Figure 3. The latter is the site where the primary reaction takes place. Both Chl-*a* and Pheophytin-*a* (Phe-*a*) are pigments known to be present in this reaction center where they act as the primary electron donor (P680) and acceptor respectively. Also present in the PSII reaction center are the plastoquinones, quinones A (QA) and B (QB) that always act in series with the carotenoid molecules. Pigments that undergo electron transfer reaction are bound to the Dl-Dz-Cytb559 protein complex, an integral part of the PSII core complex. The reaction center of PSII probably contains four molecules of Chl-*a*, two molecules of Phe-*a* and two quinones. Compared to the PSII reaction center, the PSI reaction center is more complicated. Its Chl-*a* content varies between 100 and 200 molecules, most of which are not photochemically active, but serve as part of the antenna system that absorbs light and transfers energy to the reaction centers. Excitation of PSII drives the transfer of an electron from P680 to a Phe-*a* molecule, which immediately reduces QA.

After being transferred to QB, the electron flows to the oxidized Chl-*a* in PSI (P700 through electron transport chain) provoking splitting of $H_2O$ to $O_2$ through four Manganese (Mn) atoms, which are probably bound to the proteins that hold P680. This splitting is followed by release of an electron used to reduce P680 to its original state. Same process happens in PS1 and the Chl-a in the PSI is reduced to its original states by electron from PSII. The excited electron in PSI is sent through the electron transport chain and the electron will be used to produce NADPH. During the process, electron traveling down the electron transport chain produces a chemio-static potential, which will be used by ATP synthesizer to synthesis ATP from ADP. The resulting ATP and NADPH will then be used to produce sugar from $CO_2$.

## 4. Impregnation of Chlorophyll-*a* within carboxymethyl cellulose

There are various methods to prepare *"in-vitro"* biomimetic photosynthetic films, such as spin-coating, Langmuir-Blodgett, sol-gel and electrodeposition methods. However, only spin-coating method enables true impregnation of Chl-*a* molecules due to the nature of the process to yield entrapped chromophores within the CMC matrix. When mixed with the right concentration of CMC and spin-coated onto solid substrates, a high dispersion of Chl-*a* molecules was observed. This is because cellulose derivatives are excellent dispersion agents widely used in the detergent, food, paper, textile, pharmaceutical and paint industries.

An important feature of a good dispersive agent is to provide a suitable matrix environment to the active material. For a complex photosynthetic biomolecule such as Chl-*a*, this becomes extremely crucial for optimum photoactivity to be achieved. Chl-*a* molecules embedded within the CMC and other cellulose derived polymer matrix films were able to exhibit potential in mimicking the property as photosynthetic pigment "*in-vivo*" systems. In one such work, Wrobel et al. [23] reported Chl-*a* molecules in polymer matrix of polyvinyl alcohol (PVA) and cellulose derivatives. In their work, Chl-*a* in monomeric form, dimer and nitrocellulose complex (NC) forms were investigated. The concentration of PVA used was 1 g per 9 ml dimethyl sulfoxide (DMSO) whereas concentration of NC was either 0.4 g per 5 ml DMSO (low-NC sample) or 2 g per 5 ml DMSO for high-NC sample. This results in final concentrations of Chl-*a* in films at ($1x10^{-7}$) and ($5x10^{-1}$) mol/g. Results indicated similarity of optical properties for monomeric Chl-*a* in PVA and NC by obtaining a strong α-band with a maximum at 670 nm while a small component was observed at about 720-730 nm. Results associated with optical properties confirmed that the optical spectrum was dependent on concentrations of the polymer matrix. At the higher concentration, contribution of a significant fraction of dimers shouldering around 740 nm seems to be higher compared to the absorption peak at 670 nm, indicating formation of a dimeric form of Chl-*a* molecule. A small hump at 680 nm, characteristic feature of Chl-NC complex was obtained in high-NC. Again this indicated a very slight contribution of the monomeric form as presented by the emission spectrum.

The present work looks into the possibility of employing CMC as the AM material and investigates its localized nano-environment for the impregnation of Ch-*a* pigments and optimum photosensitivity. Following sections will provide information on materials preparation followed by experimental results and in-depth discussions. The last section meanwhile would state major conclusions and findings of the research.

## 5. Optimum CMC concentration effect

For the current work, a target Chl-*a* suspension in chloroform ($CHCl_3$) was prepared as suggested in literature [22, 23]. A concentration of 0.45 mg/ml was prepared and stored in a fridge for three days prior to usage. Seven concentrations of CMC gel-like aqueous solution (6, 12, 18, 24, 30, 36 and 41 mg/ml) were prepared for the purpose of investigating productive photoactivity based upon optimum concentration of AM concentration. From the method employed in literature [33], CMC was dissolved in deionised (DI) water and ethanol according to its solvent ratio in miligrams as shown in Table 1. The CMC solutions were then ultrasonificated for an hour and magnetic stirred at 70°C overnight. Solutions obtained appeared transparent and did not precipitate from the solvent.

| Solvent ratio (DI:Ethanol:CMC), *mg* | Mass of CMC, *mg* (±0.1) | Volume of DI, *ml* (±0.1) | Volume of Ethanol, *ml* (±0.1) | Concentration, *mg/ml* |
|---|---|---|---|---|
| 120:38:1 | 50 | 6.0 | 2.41 | 6 |
| 60:19:1 | 100 | 6.0 | 2.41 | 12 |
| 42:12:1 | 50 | 2.1 | 0.76 | 18 |
| 30:9.5:1 | 200 | 6.0 | 2.41 | 24 |
| 25:7:1 | 85 | 2.1 | 0.76 | 30 |
| 20:6:1 | 300 | 6.0 | 2.41 | 36 |
| 18:5:1 | 100 | 1.8 | 0.63 | 41 |

**Table 1.** Preparation of different concentrations of CMC suspension.

In order to investigate possible enhanced properties of CMC-Chl-*a* films prepared at different spin rates, its corresponding optical spectra were obtained. As seen from the graphs in Figure 4, similarity of absorption peaks of Soret and $\alpha$-bands at 434 and 675 nm respectively were observed. It confirms retention of the native spectroscopic properties of Chl-*a* preserved within the CMC matrix. In certain CMC concentrations (Figure 4(d) and (g)), higher absorptions were registered at both bands compared to free Chl-*a* molecules suggesting applications in "in-vitro" biomimetic immobilizing media in synthetic mixture [34]. Elsewhere [22, 23], higher values of absorbance peaks have been reported.

However in the present work, intense absorption at $\alpha$-band were obtained when Chl-*a* molecules were impregnated into the CMC matrix with concentration of 24 mg/ml and especially at 1000 rpm (Figure 5). At this concentration and rpm, the optimum dispersive concentration for Chl-*a* molecules possibly allows enhanced photoreaction to occur when evenly distributed in polymer hydrogel forms. Higher photosensitivity was also observed for the other films when prepared at the optimum spin rate. Careful observation of the graphs in Figure 5 also demonstrated a sharp and narrow absorption peak at 415 nm of 10 nm width attributed to the Soret band instead of the usual 435 nm in free Chl-*a* molecules. Upon increase in matrix concentration, the characteristic peak observed was observed to generally shift to a higher angle. The resulting blue shift may be due to even existence of Chl-*a* molecules rather than being aggregated within the CMC support. This explains the higher depth values obtained at CMC concentrations of 18 and 30 mg/ml as compared to other concentrations. Since absorption intensity is also dependent on film thickness, the resulting decrease in absorption peaks of CMC-Chl-*a* films of increasing spin rates is quite understandable. Rougher surfaces register lower photosensitivity even at the optimum CMC concentration. This may be due to spin rate influencing colloid present within the film environment besides the concentration factor.

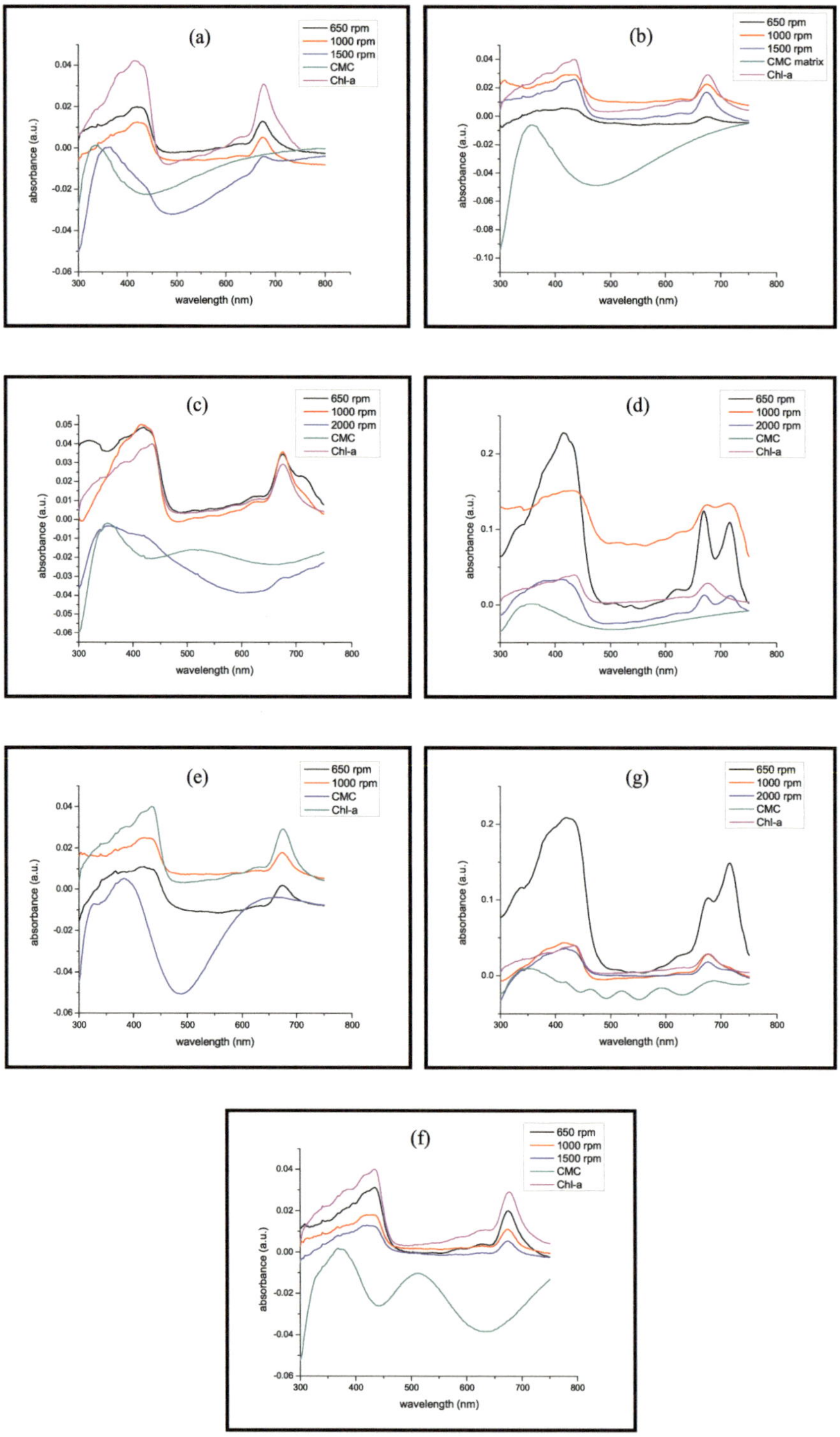


**Figure 4.** CMC-Chl-*a* thin films at concentrations of (a) 6, (b) 12, (c) 18, (d) 24, (e) 30 and (f) 41 mg/ml of CMC matrix respectively.

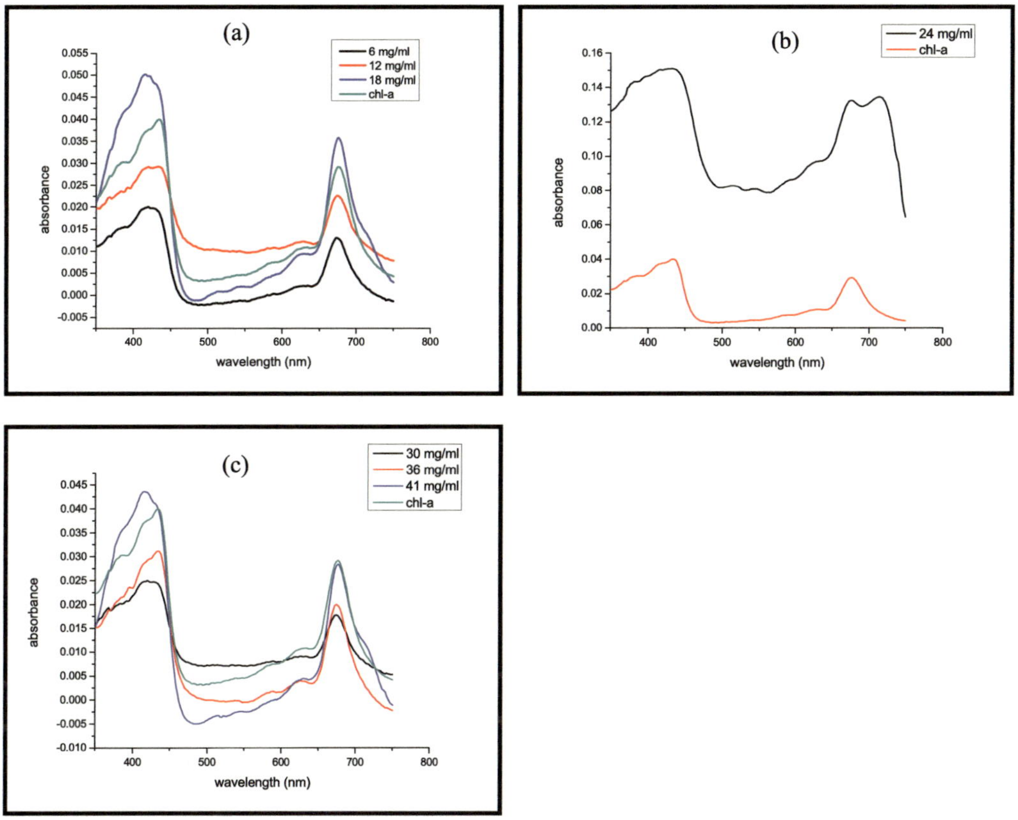


**Figure 5.** (a) Absorption spectra of Chl-*a* in CMC matrix films and Chl-*a* ($5 \times 10^{-4}$ M) in chloroform for CMC matrix concentration below 24 mg/ml. (b) A similar spectrum has been obtained for Chl-*a* (at the same pigment concentration) in CMC matrix of concentration 24 mg/ml showing a much higher peak (roughly 3 times) and in (c) CMC matrix of concentration greater than 24 mg/ml.

The optimum CMC concentration at 24 mg/ml with the most intense absorption at $\alpha$-band registered the largest full width half maximum contributed by a significant fraction of dimers shouldering around 720 nm. High solvent content also provides an aqueous media for photoreaction of Chl-*a* molecules resulting in a loose CMC structure. This in turn provides a suitable nano-environment for the impregnation of the photosynthetic pigments. Such an environment allows uniform distribution and dispersion of the molecules within the synthetic medium for enhanced photosensitivity. Higher solvent volume reflects higher concentration of CMC aqueous media, in agreement with literature [23] stating influence in the absorption spectra of Chl-*a* films. Therefore, the native state of Chl-*a* molecules were still retained in cellulose derivatives even in high concentrations of CMC. The work also suggests self-association of Chl-*a* molecules by means of a hydrogen-bonding nucleophile, such as water or solvent. Higher absorption peak intensity values were obtained at the optimum concentrations; comparable to previous works elsewhere [22]. The graphs in Figure 6 show optical absorbance of CMC-Chl-*a* films over time for CMC matrix concentrations of 18, 24 and 41 mg/ml, respectively. In this experiment, spectroscopic characterization was carried-out every week for a month for the films with higher α-band peaks compared to free Chl-*a* molecules in chloroform.. Studying absorbance peaks at the α-band indicates photophysical functionality of Chl-*a* molecules to synthetic mixtures. Encapsulating the pigments with the CMC layer slows down oxygen reduction process from oxidizing the Chl-*a* molecule in the films prepared. As a result, retention of the photosynthetic nature of the reconstituted Chl-*a* molecules are maintained successfully.

In a previous work [35], researchers have reported loss of spectroscopic properties when Chl-*a* molecules deposited as thin films were exposed directly to oxygen. As such, we have shown that by impregnating Chl-*a* molecules into CMC matrix forming thin films, an encapsulating effect is achieved which prolongs the optical functionality of the photosynthetic pigment. Eventual decay in photosensitivity may be attributed to the fact that some molecules at air-solid interface may not be totally encapsulated against the elements due to the nature of the process. In addition, CMC-Chl-*a* films also exhibit a small blue shift of the optical absorbance from 674 to 672 nm, indicative of gradual structural changes in the Chl-*a* molecules. At CMC concentrations of 18, 24 and 41 mg/ml, the films are more likely to have loose matrix structure in its hydrogel form. This results in localized nano-environment for suitable entrapment of Chl-*a* molecules for optimum photoactivity indicated by high value of absorption peak intensity at the α–band.

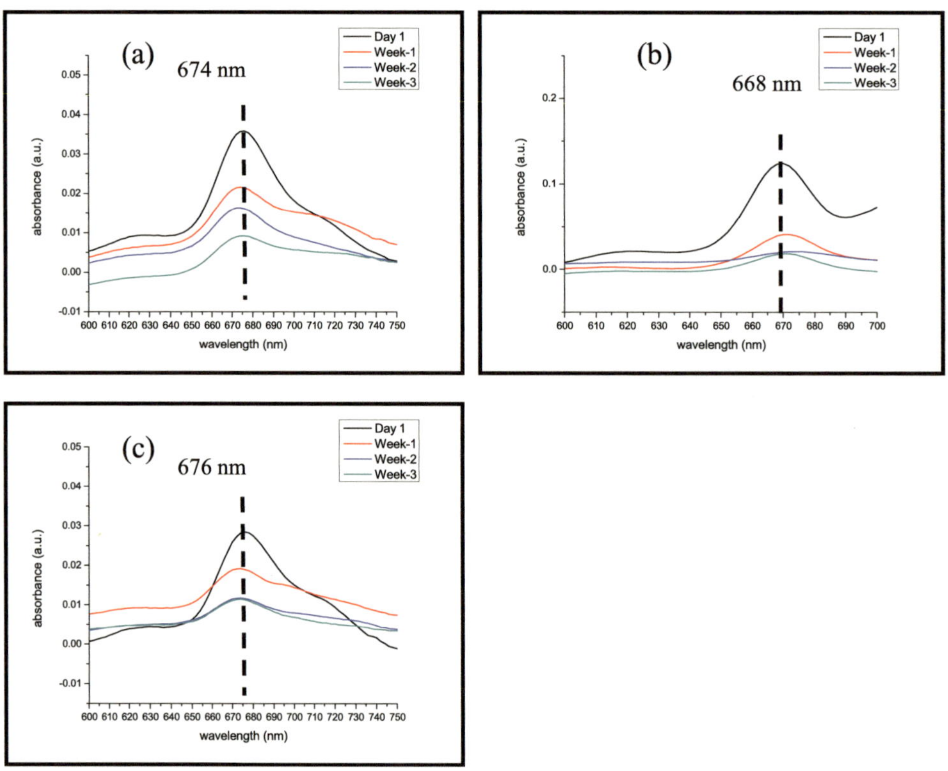


**Figure 6.** Absorption spectra at α-band for CMC concentrations of (a) 18, (b) 24 and (c) 41 mg/ml respectively.

## 6. Structural properties of CMC-Chl-*a* films

Figure 7 shows XRD measurement of the CMC-Chl-*a* films prepared at spin rate of 1000 rpm and different concentrations. Regardless of the spin rate effects, similarity of XRD patterns for all matrix concentrations in the range of 2θ between 10° to 80°, confirms existence of Magnesium nitrate di(N-N-dimethylurea) tetrahydrate and Magnesium nitrate-6-ethanol identified as crystalline phase.

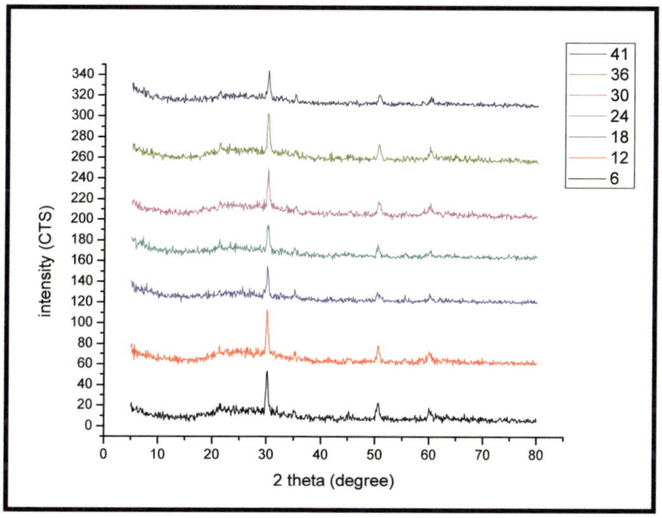


**Figure 7.** XRD patterns of CMC-Chl-*a* films prepared at 1000 rpm and different CMC concentrations.

In Figure 8, the most intense peak at 2θ occurs at 30.1° referring to the Chl-*a* characteristic peak which was shifted to higher angle when films were prepared in higher concentrations. Broadening of diffraction lines was also observed when films were prepared in the optimum CMC concentration of 24 mg/ml resulting in expansion of pores in the films [36]. This also indicates the existence of a smaller unit cell being the dominant effect causing the breadth of the diffraction lines [37]. Based on XRD results showing several sharp peaks, the polycrystalline nature of Chl-*a* molecules impregnated within the organic polymer matrix could be confirmed. Using Scherer equation, the grain size of a single Chl-*a* molecule is calculated to be about 10 nm of length confirming existence of nanocrystalline structures of the photosynthetic pigments.

Figure 9 represents the FESEM-EDX imaging measurement, which is in good agreement with the XRD measurement. Existence of elements such as magnesium oxide, sodium, oxygen and indium arsenide elements are traced within the prepared colloids. Since CMC may interact with Chl-*a* molecules and influence morphology of the films formed on ITO substrates, AFM investigation was performed to obtain morphological information of Chl-*a* embedded into the CMC polymer matrix. Figure 10 is AFM images showing morphology of CMC-Chl-*a* colloids at CMC matrix concentration of 24 mg/ml deposited on ITO slides at the spin rate of 1000 rpm. Generally, the colloids were in long and unfolded successions of aggregates, regardless the concentrations of the matrices possibly due to different colloidal interaction with the substrate.

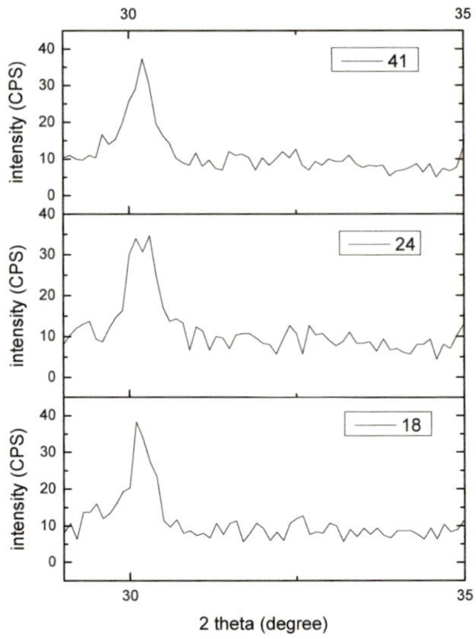


**Figure 8.** XRD patterns of the films in the region of 2θ = 29° - 35° for Chl-*a* embedded in different concentrations of CMC matrix: (a) 18, (b) 24 and (c) 41 mg/ml.

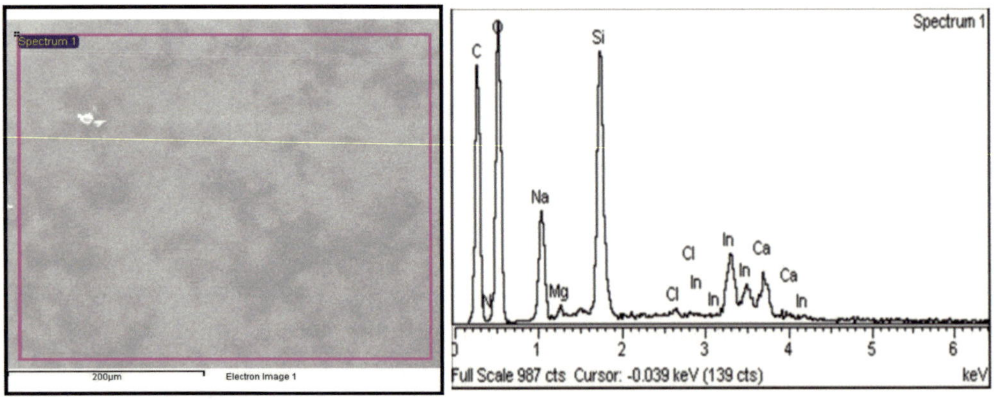


**Figure 9.** FESEM-EDX analysis of CMC-Chl-*a* thin film prepared using 24 mg/ml of CMC aqueous solution. EDX images were obtained under 300X magnification with accelerating voltage of 20 kV within a processing time of 5 seconds.

According to Boussaad *et al.* [36], CMC-Chl-*a* complex in solution with polar solvents usually carries a net positive charge. As a consequence, complex-substrate interaction would be influenced by repulsive or attractive interaction as a result of the presence of this charge. The repulsive interaction induced by the ITO surface may cause the colloids of the complex to unfold into long successions of aggregates. It was also clear from literature that long and unfolded successions of aggregates were observed when the complex is deposited onto hydrophilic surfaces, such as gold [38]. Results show that different concentrations of CMC aqueous solutions used to prepare thin films at certain spin rates could produce different degrees of orderly distribution of CMC-Chl-*a* complex across the substrate. Resulting thin films as shown in Figure 11 therefore show differing surface roughness. In these films,

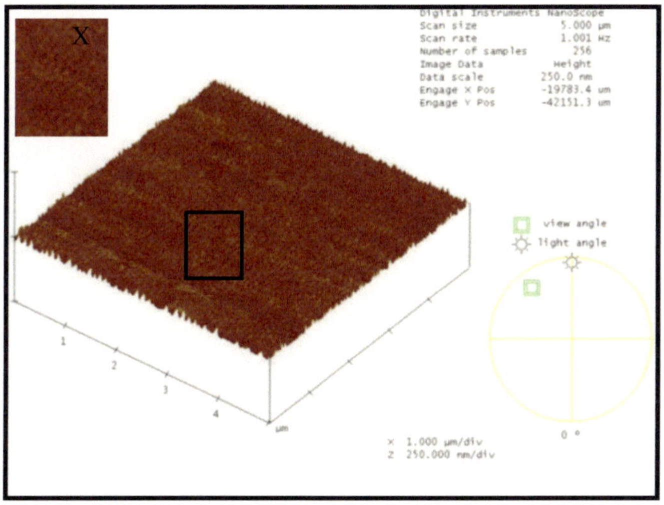


**Figure 10.** AFM images of CMC-Chl-*a* films deposited onto ITO slides with CMC concentration of 24 mg/ml. Micrograph X was cropped and enlarged from the square box in the AFM image.

crescent shaped Chl-*a* colloids of length of 10 nm are observed throughout the substrate. Such uniformity of molecular structures was observed to decrease at high concentrations, an observation in good agreement with the XRD results. This may also explain the formation of dimer association in the aggregates. Arrangement of dimers within the nanocrystal reflects the crystalline structure of the microcrystalline Chl-*a*. The differences between dimer and microcrystal structures of Chl-*a* in the films depend strongly on the dissociation and association of the dimers. Usually, excess of solvent used in the CMC matrix compete with the nanocrystal bonding and causes the aggregates to dissociate resulting in rougher surfaces. However, angular velocity of the spinning method may also influence the formation of dimer association in the aggregation [21].

Figure 12 summarize the roughness (a) and depth analysis (b) respectively for CMC-Chl-*a* films at different spin rates. In general, from Figure 12(a), it could be deduced that Chl-*a* molecules were well dispersed in the CMC matrix. This results in surface mean roughness in the range of 4 to 19 nm, quite comparable to other deposition techniques [22, 23]. The values of roughness were found to decrease as the CMC concentrations increases. Roughness meanwhile increases after achieving the optimum concentrations for each spin rates. Optimum concentration of CMC was achieved at 36 mg/ml for the spin rate of 650 rpm which results in mean roughness of about 3.86 nm. At spin rates of 1000 and 1500 rpm, optimum concentration was achieved at 18 and 30 mg/ml respectively. Hence, it could be assumed that mean roughness values of the CMC-Chl-*a* films were independent to spin rate. Higher spin rates did not produce a smoother film except for the 30 mg/ml concentration of CMC. The values actually decreased when films are spun at higher spin rates.

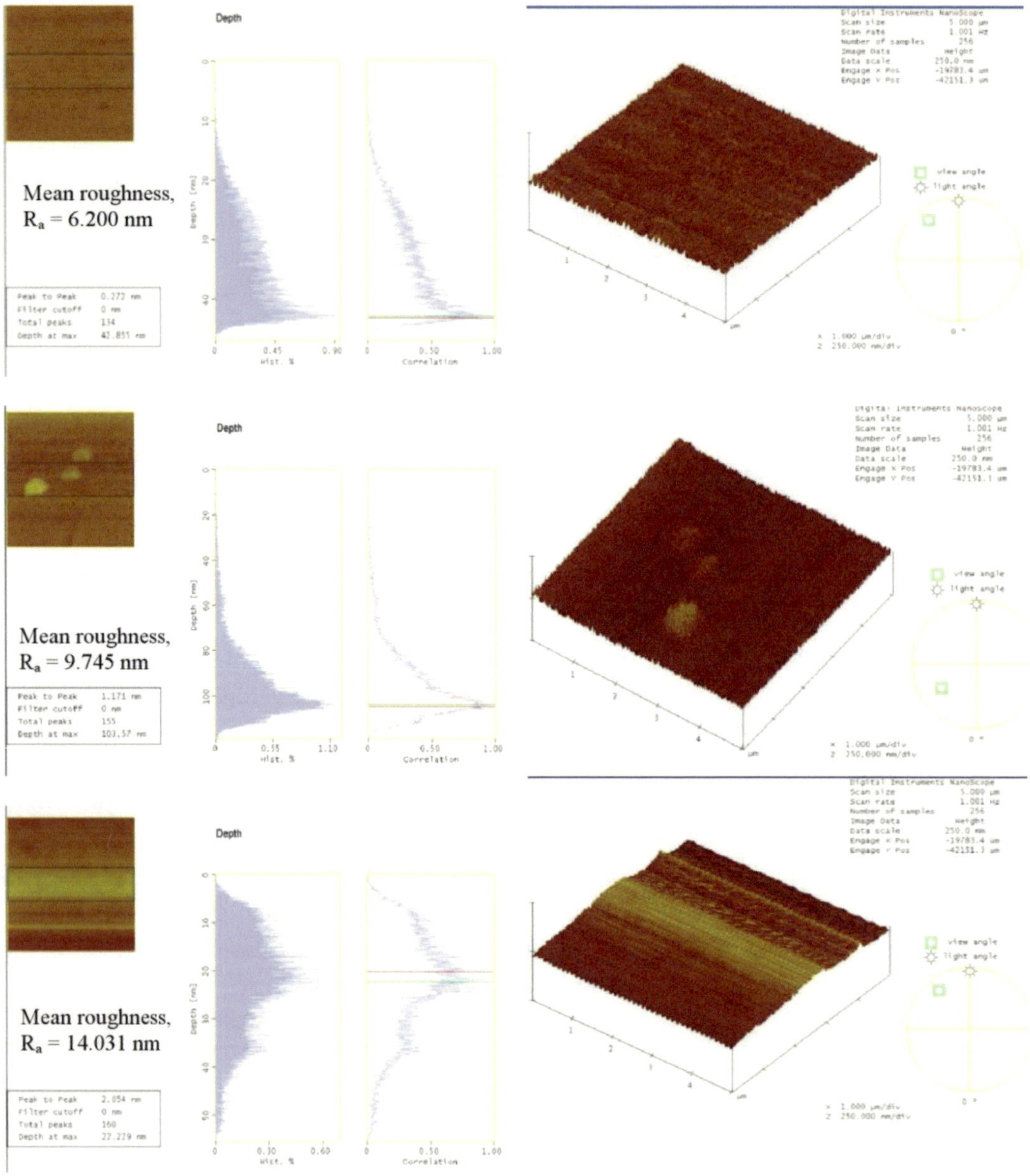


**Figure 11.** AFM images of CMC-Chl-*a* thin films prepared at spin rates of 650, 1000 and 1500 rpm at CMC concentration of 24 mg/ml.

Results of depth analysis for CMC-Chl-*a* films indicate that the prepared concentrations of CMC aqueous solution were able to produce better distribution of Chl-*a* molecules across the substrate. As shown in Figure 12(b), this results in maximum depth in the range between 17 to 100 nm. As observed from the roughness analysis results, depth results also confirm that the films prepared at the spin rate of 1000 rpm shows the smallest deviation between the values. The values increased with spin rates resulting in thicker films due to dissociation

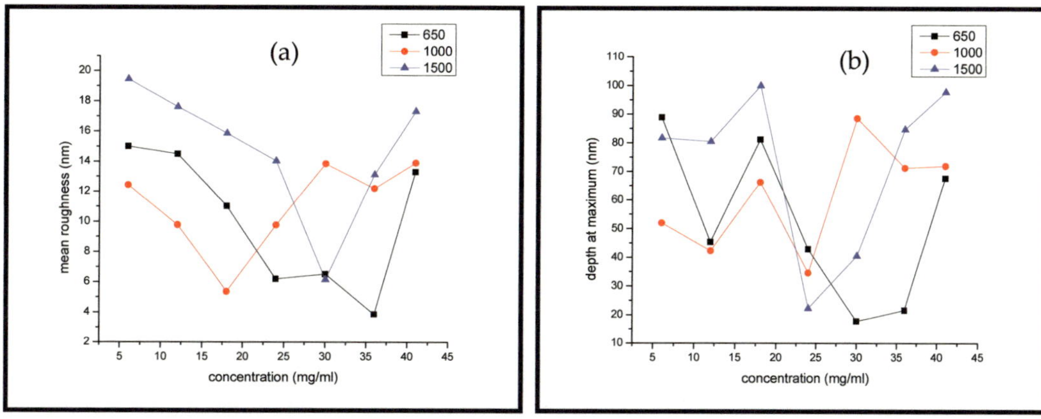


**Figure 12.** (a) Mean roughness and (b) depth analysis of CMC-Chl-*a* films.

of Chl-*a* aggregates within CMC matrix. Therefore, weak affinity between Chl-*a* colloids and the ITO substrate could be deduced, as reported in literature [13]. When weakly binding onto ITO surfaces, solvent bound molecules were found harder to be controlled by spin rate and thus create thicker films. However, for CMC-Chl-*a* films prepared in optimum matrix concentration of 24 mg/ml, the values were observed to decrease with increase in spin rates.

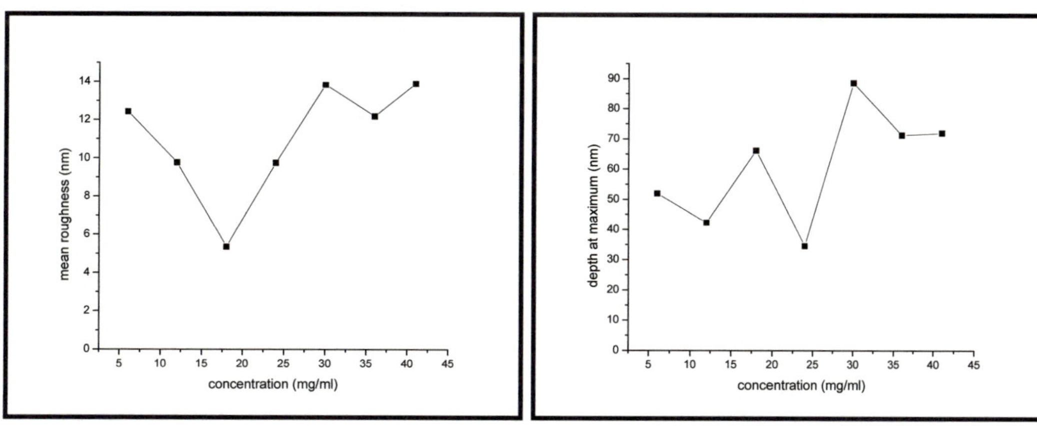


**Figure 13.** (a) Mean roughness and (b) depth analysis of CMC-Chl-*a* films at spin rate of 1000 rpm.

In summary, the Chl-*a* molecules were well dispersed within the cellulose derivative while achieving desirable thickness and smooth thin films regardless of spin rate. AFM images also confirmed that the spin coating technique could be employed to synthesize nanocrystalline molecular structures of Chl-*a* molecules at different concentrations of CMC matrices. Particularly, CMC concentration of 24 mg/ml exhibits the optimum AM concentrations at the spin rate of 1000 rpm as shown in Figure 13(a). Depth analysis meanwhile in Figure 13(b) concludes molecular structures smaller than 70 nm at concentrations below 24 mg/ml. These findings indicate strong affinity between high concentrations of CMC and ITO substrate. Upon strongly binding to ITO surface, solvent bound molecules were found easier to be controlled by the spin rate and thus create a new surface for the CMC-Chl-*a* film. This new surface might bring about the formation of smoother and thinner film morphology. However, excess solvent used competes with nanocrystal bondings causing the aggregates to dissociate resulting in rougher surfaces.

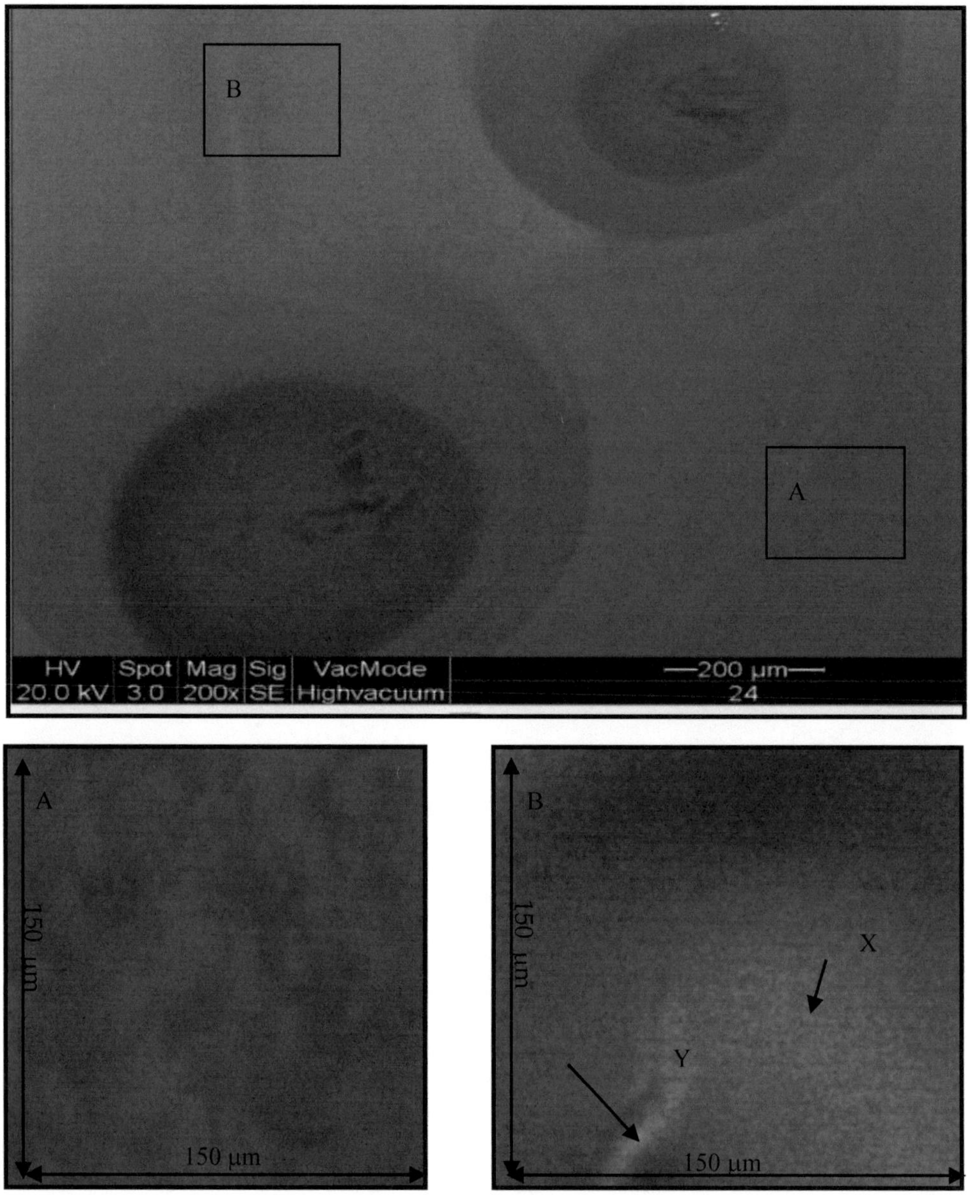


**Figure 14.** FESEM views of spin-coated ITO for CMC-Chl-*a* films. Micrograph A and B were cropped and enlarged three times from the main micrograph

FESEM measurements were also performed to further confirm existence of nanocrystal form of Chl-*a* molecules in CMC matrices at the optimum concentration and spin rate. Figure 14 shows the micrographs obtained for CMC concentration of 24 mg/ml at 1000 rpm. In a separate work [11] involving dry CMC films on pyrolytic graphite electrodes, it was reported that large amount of water absorbed forming hydrogel by obtaining a wide leaf-like structure without formation of crystals. However, the SEM top views of CMC-Chl-*a* films in the present study show a different surface morphology. The aggregates were compressed within the colloid resembling large cloud-like thin sheets and form a very dense complex on the ITO surface (micrograph A). Micrograph B meanwhile shows orderly distributed molecular structures highlighted by the spots within the square box B in the main micrograph of Figure 14. These spots were distributed randomly across the substrate indicative of the uniformity of the fabricated films, confirming arrangement of dimers within the nanocrystalline structures found in the colloids (X). Organization of Chl-*a* molecules reflects the optical properties of the nanocrystalline molecular structure. Some small "island-like" formation in highly packed groups (Y) found in the colloids (bright spots) is actually Chl-*a* dimers [23] embedded into the CMC matrix forming tight interactions. They are responsible for successful retention of Chl-*a* molecules within the films and provide a "membrane-like" environment for entrapment and enhanced functionality of the photosynthetic pigments.

## 7. General conclusions

This study suggests the self-association of Chl-*a* in cellulose derivative matrices utilizing the spin-coating technique while maintaining its native spectroscopic characteristics when reconstituted into CMC matrix. It highlights the potential of encapsulating Chl-*a* and other PBMs with the CMC acting as the AM. Successful prolonged retention of the Chl-*a* molecules' spectroscopic properties within the CMC-Chl-*a* films was also demonstrated. Considerably smooth film surfaces were obtained comparable to other conventional deposition techniques.

High absorption peak intensities of the Soret and $\alpha$-bands indicate well-dispersed Chl-*a* molecules within the CMC phase. Lower solvent content or lower concentration effect causes the Chl-*a* aggregation to exhibit weak affinity to the ITO substrate. As a consequence, it becomes harder to control film morphology at high spin rates resulting in thicker films. Comparable results were also deduced from XRD patterns highlighting retention of the native state of Chl-*a* molecules for all CMC matrix concentrations.

Morphological characterization confirms the nanocrystal formation of Chl-*a* molecules in all CMC concentrations of the synthetic film. The colloids were observed to be in long and unfolded successions of aggregates indicating self-association of Chl-*a* molecules in the cellulose derivative matrix. Maximum depth was found to be in the order of 100 nm and would further decrease if the films were prepared in optimum CMC concentration of 24 mg/ml. This may well suggest dimer association in the aggregation by further observing the absorption band at 720 nm. The arrangement of dimers within the nanocrystal in turn reflects the crystalline structure of the overall microcrystalline Chl-*a* structure in the films prepared.

## Author details

P. Vengadesh
*University of Malaya, Malaysia*

## Acknowledgement

The experimental section was supported in parts by the Ministry of Higher Education, Fundamental Research Grant Scheme (FRGS) grant number FP011/2008C, University of Malaya Postgraduate Research Fund grant number IPPP/UPDit/Geran(PPP)/PS225/2009A and High Impact Research (HIR) grant number J-21002-7-3823. The author is also grateful to his postgraduate student, Wong Yee Wei who was reponsible for the data collection.

## 8. References

[1] Ioana L, Nicoleta B, Rodica N, Alina M, Florin M, Ion I, Aurelia M (2010) Encapsulation of Fluorescence Vegetable Extracts within a Template Sol-Gel Matrix. optical materials 32: 711-718.

[2] Zhaoqi F, Chuanhui C, Shukun Y, Kaiqi Y, Rensheng S, Daocheng X, Chunyu M, Xu W, Yuchun C, Guotong D (2009) Red and Near-Infrared Electroluminescence from Organic Light-Emitting Devices based on a Soluble Substituted Metal-Free Phthalocyanine. optical materials 31: 889-894.

[3] Wong YW, Vengadesh P (2011) Synthesis, Structural and Spectroscopic Properties of Encapsulated Chlorophyll-*a* Thin Film in Carboxymethyl Cellulose. j. of porpyrin and phythalocyanine 15: 122-130.

[4] Itamar W, Shai R (2003) Control of the Structure and Functions of Biomaterials by Light. ang. chemie int. ed. in eng. 35(4): 367-385.

[5] Mohamed El-Anwar HO, Fatma E (1988) Photosynthetic Electron Transport under Phosphorylating Conditions as Influenced by Different Concentrations of Various Salts. j. exp. bot. 39(7): 859-863.

[6] Devens G, Thomas M (2002) Active Transport of $Ca^{2+}$ by an Artificial Photosynthetic Membrane. nature 420(6914): 398-401.

[7] Hemadeh O, Chilukuri S, Bonet V, Hussein S, Chaudry IH (1993) Prevention of Peritoneal Adhesions by Administration of Sodium Carboxymethyl Cellulose and Oral Vitamin E. surgery 114(5): 907-10.

[8] Lin D, Zhao Y (2007) Innovations in the Development and Application of Edible Coatings for Fresh and Minimally Processed Fruits and Vegetables. In: Comprehensive Reviews in Food Science and Food Safety. Oregon: Institute of Food Technologists.

[9] Minami N, Kim Y, Miyashita K, Kazaoui S, Nalini B (2006) Cellulose Derivatives as Excellent Dispersants for Single-Wall Carbon Nanotubes as Demonstrated by Absorption and Photoluminescence Spectroscopy. applied physics letters 88(9): 093123-093126.

[10] He Huang PH (2003) Electrochemical and Electrocatalytic Properties of Myoglobin and Hemoglobin Incorporated in Carboxymethyl Cellulose Films. bioelectrochemistry 61: 29-38.

[11] Nisperos-Carriedo MO (1994) Edible Films and Coatings based on Polysaccharides. In: Krochta JM, Baldwin EA, Nisperos-Carriedo MO, editors. Edible Coatings and Films to Improve Food Quality. Lancaster: Technomic Publishing Company Inc. pp. 305–35.

[12] Carboxymethyl Cellulose.Retrieved from: http://www.Isbu.ac.uk/water/hycmc.html
[13] Bourtoom T (2008) Edible Films and Coatings: Characteristics and Properties. journal of international food research. 15(3): 237-248.
[14] Savage AB, Young AE, Maasberg AT (1954) Cellulose and Cellulose Derivatives. In: Ott E, Sourlin HM, Grafflin MW, editors. High Polymers. New York: Interscience.
[15] Nelson KL, Fennema OR (1991 Methylcellulose Film to Prevent Lipid Migration in Confectionery Products. j. food sci. 56: 504–509.
[16] Gennadios A, Hanna MA, Kurth B (1997) Application of Edible Coatings on Meats, Poultry and Seafoods: A Review. lebens wissen technol. 30: 337–350.
[17] Kaistner U, Hoffmann H, Donges R, Hilbig J (1997) Structure and Solution Properties of Sodium Carboxymethyl Cellulose. colloids surf. A 123: 307–328.
[18] Jacobs EE, Vatter A E, Holt AS (1953) Crystalline chlorophyll and bacteriochlorophyll. journal of chemistry and physics 21: 2246-2279.
[19] Katz JJ (1994) Long wavelength chlorophyll. spectrum 7(1): 1, 3-9.
[20] Oureriagli A, Kass, H, Hotchandani S, Leblanc RM (1992) Analysis of Dark Current-Voltage Characteristics of Al/Chlorophyll a/Ag Sandwich Cells. journal of applied physics 71(11): 5523-5530.
[21] Kassi H, Hotchandani S, Leblanc RM (1993) Hole Transport in Microcrystalline Chlorophyll a. appl. phys. lett. 62(18): 2283-2285.
[22] Pasquale LD, Lucia C, Pinalysa C, Paola F and Angela A (2004) Photophysical and Electrochemical Properties of Chlorophyll a-Cyclodextrins Complexes. bioelectrochemistry 63: 117– 120.
[23] Wróbel D, Planner A. and Perska B. (1996) Time-Resolved Delayed Luminescence of Chlorophyll $\alpha$ in Anhydrous Polymer Systems. spectrochimica acta part a 52: 97-105.
[24] Chlorophyll-a. Retrieved from: http://en.wikipedia.org/wiki/Chlorophyll
[25] Mehraban Z, Farzaneh F, Shafiekhani A (2007) Synthesis and Characterization of a New Organic-Inorganic Hybrid NIO-Chlorophyll-*a* as Optical Materials. optical materials 29: 927-931.
[26] Photosystem II. Retrieved from: http://beckysroom.tripod.com/summary1.htm
[27] Rodrigo FA (1953) Preliminary Note on Experiments Concerning the State of Chlorophyll in the Plant. biochim. biophys. acta 10(2): 342.
[28] Jacobs EE, Holt AS, Rabinowitch E (1954) The Absorption Spectra of Monomolecular Layers of Chlorophyll *a* and Ethyl Chlorophyllide *a*. j. chem. phys. 22: 142-144.
[29] Ke B (1966) In: Vernon LP, Seely GR, editors. The Chlorophylls. New York: Academic Press Inc. pp. 253.
[30] Chapados C, Leblanc RM (1977) Aggregation of Chlorophylls in Monolayers. Infrared Study of Chlorophyll a in a Mono- and Multilayer Arrays. chem. phys. lett. 49: 180-182.
[31] Leblanc RM, Chapados C (1977) Aggregation of Chlorophylls in Monolayers. biophys. chem. 6: 77-85.
[32] Chapados C, German D, Leblanc RM (1980) Aggregation of Chlorophylls in Monolayers: IV. The Reorganization of Chlorophyll a in Multilayer Array. biophys. chem. 12: 189-198.

[33] Jagadeesh Babu Veluru SK (2007) Electrical Properties of Electrospun Fibers of PANI-PMMA Composites. journal of engineered fibers and fabrics 2(2): 25-31.

[34] Project Ingenhousz: Light, Genes and Molecular Machines. Retrieved from http://dwb.unl.edu/Teacher/NSF/C11/C11Links/photoscience.la.asu.edu (Accessed 30/04/99)

[35] Mapel JK (2006) The Application of Photosynthetic Materials and Architectures to Solar Cells. California: Massachusetts Institute of Technology.

[36] Boussaad S, Tazi A, Leblanc RM (1999) Study of the Conformational Changes of Chlorophyll a (Chl a) Colloids with the Atomic Force Microscope. j. of colloid and interface science 209: 341-346.

[37] Cullity B (1956) Elements of X-Ray Diffraction. USA: Addison-Wesley Publishing Company Inc.

[38] Dharmadhikari CV, Ali AO, Suresh N, Phase DM, Chaudhari SM, Gupta A, Dasannacharya BA (2000) A Comparison of Nucleation and Growth Investigations of Thin Films using Scanning Tunneling Microscopy, Atomic Force Microscopy and X-ray Scattering. materials science and engineering b 75(1): 29-37.

Chapter 4

# Optical, Magnetic, and Structural Properties of Semiconductor and Semimagnetic Nanocrystals

Ricardo Souza da Silva, Ernesto Soares de Freitas Neto and Noelio Oliveira Dantas

Additional information is available at the end of the chapter

http://dx.doi.org/10.5772/46037

## 1. Introduction

Semiconductor and semimagnetic nanocrystals (NCs), grown in different host materials, have attracted considerable attention due to their unique properties which are caused by zero-dimensional quantum confinement effects. Several advances in controlled chemical synthesis of materials have provided ways to grow NCs and manipulate their size, shape, and composition using different methodologies [1-5]. The interesting properties of these nanoparticles can be explored in diverse technological applications, such as wavelength tunable lasers, light-emitting devices, solar cells, and spintronic devices among others [6-13]. A detailed and comprehensive understanding on the properties of these NCs should be achieved in order to target many of the possible technological applications.In this chapter, we will present our main results and discussions on the optical, magnetic, and structural properties of semiconductor and semimagnetic NCs that were successfully grown by the melting-nucleation approach or by the chemical precipitation method.

Optical processes in PbS NCs were investigated by employing the following experimental techniques: optical absorption (OA), photoluminescence (PL) and atomic force microscopy (AFM). The OA and PL peaks of these PbS NC samples showed a separation of about 0.05-0.20 eV, confirming thus the large Stokes shift. A comprehensive understanding on this large Stokes shift was achieved by investigating the radiative and nonradiative processes in these nanoparticles [14].

We will report evidences to the induced migration of $Mn^{2+}$ ions in $Cd_{1-x}Mn_xS$ NCs by selecting a specific thermal treatment to each NC sample. The characterization of these magnetic dots was investigated by the electronic paramagnetic resonance (EPR) technique. The comparison of experimental and simulation of EPR spectra confirms the incorporation of $Mn^{2+}$ ions both in the core and at the dot surface regions. The thermal treatment to a magnetic sample, via selected annealing temperature and/or time, affects the fine and hyperfine interaction constants which modifies the shape and the intensity of an EPR transition spectrum. The identification of these changes has allowed tracing the magnetic ion migration from core to surface regions of a dot as well as inferring on the local density of the magnetic impurity ions [15].

The properties of $Pb_{1-x}Mn_xS$ NCs embedded in a borosilicate glass matrix has been investigated by magnetic measurements. The data indicated that only a small fraction of the nominal Mn-doping was incorporated into the PbS NCs, in both 0.3% and 0.7% nominal doping ends. Moreover, low temperature magnetization and susceptibility data showed that most of the magnetic ions hosted by the of $Pb_{1-x}Mn_xS$ NCs are in a paramagnetic state[16, 25].

We also have employed the magnetic force microscopy (MFM) in order to study the magnetic moments of Mn-doped nanoparticles, namely: $Cd_{1-x}Mn_xS$ and $Pb_{1-x}Mn_xS$ NCs. In these measurements, the interaction between tip and NC magnetization induces the contrast observed in the MFM images. A dark area (light area) in this contrast is caused by attraction (repulsion) between tip and NC magnetization. Evidently, the magnetization in each NC is caused by the size-dependent *sp-d* exchange interactions, proving that $Mn^{2+}$ ions are actually incorporated into the semimagnetic nanostructures. Therefore, all these results certainly demonstrate that MFM is a powerful technique that plays a very important role in order to investigate semimagnetic nanocrystals [4, 17].

$Zn_{1-x}Mn_xO$ NCs were successfully grown by the chemical precipitation method and their magnetic properties were effectively investigated by the EPR technique. Thus, we have confirmed the actual incorporation of $Mn^{2+}$ ions into the hosting ZnO NCs, while the hexagonal wurtzite structure of these nanoparticles was preserved. The well known $Mn^{2+}$ six hyperfine lines in the EPR spectra of the as-produced samples were clearly observed. In addition, as the Mn-concentration increases to a level of about 0.81% a broad EPR line is observed, thus confirming the onset of Mn-Mn exchange interaction [5].

The structural properties of these $Zn_{1-x}Mn_xO$ NCs were characterized by Raman spectroscopy and X-Ray Diffraction (XRD) measurements. The observed shift in the diffraction peaks toward lower angles, with increasing in the x-concentration, was attributed to incorporation of $Mn^{2+}$ ions into the ZnO NCs. This analysis is strongly corroborated by results obtained by the Raman spectroscopy, where the data have also provided evidences of the replacement of zinc ions by manganese ions into the $Zn_{1-x}Mn_xO$ NCs. Besides the Raman features typical of the ZnO structure, the $Zn_{1-x}Mn_xO$ ($x > 0$) nanoparticles display an extra Raman peak at 659 $cm^{-1}$. This finding is a strong evidence of the replacement of zinc ions by manganese ions [5, 18].

The results of this chapter confirm the high quality of the semiconductor and semimagnetic NCs that were successfully grown by the melting-nucleation approach or by the chemical precipitation method. The comprehensive discussions that were presented on the properties of nanoparticles certainly demonstrate the great potential of these systems for various technological applications. We believe that this chapter can motivate further investigations and applications of other systems containing NCs.

## 2. Synthesis of nanocrystals

The development of nanocrystals (NCs) produced of controlled way for possible applications technologic, depends on the synthesis methodology adopted. We report the study of $Pb_{1-x}Mn_xS$ and $Cd_{1-x}Mn_xS$ NCs synthesized in borosilicate glass matrix template using the fusion method and $Zn_{1-x}Mn_xO$ NCs using the co-precipitation method.

### 2.1. Synthesis of $Pb_{1-x}Mn_xS$ nanocrystals

$Pb_{1-x}Mn_xS$ NCs were produced by the fusion method in the glass matrix with the following nominal composition: $40SiO_2 \bullet 30Na_2CO_3 \bullet 1Al_2O_3 \bullet 25B_2O_3 \bullet 4PbO$ (%mol), herein quoted as SNABP glass matrix. The nominal composition of the nanocomposite was achieved by adding 2S (%wt) plus $x$Mn with respect the (1-$x$)Pb, with $x$ = 0, 0.003 and 0.007. The samples were produced following two major preparation steps. In the first step the powder mixture was melted in an alumina crucible at 1200°C for 30 minutes, following a quick cooling of the crucible containing the melted mixture from 1200°C down to room-temperature. At the end of this step a first series of samples labeled SNABP: $x$Mn were produced for further characterization. In the second step thermal annealing of the previously-melted glass matrix (SNABP: $x$Mn samples) was carried out at 500°C for times different, with the purpose to enhance the diffusion of $Pb^{2+}$, $Mn^{2+}$, and $S^{2-}$ species within the hosting matrix. Due to the thermal annealing procedure $Pb_{1-x}Mn_xS$ NCs were formed within the glass template [7].

### 2.2. Synthesis of $Cd_{1-x}Mn_xS$ nanocrystals

$Cd_{1-x}Mn_xS$ NCs were synthesized in a glass matrix (SNAB) with a nominal composition of $40SiO_2$ . $30Na_2CO_3$ . $1Al_2O_3$ . $29B_2O_3$ (%mol) + 2[CdO + S] (%wt), and Mn-doping concentration (x) varying with respect to Cd-content from 0 to 10%. The first step of sample preparation consisted of melting powder mixtures in an alumina crucible at 1200°C for 30 minutes. Then, the crucible containing the melted mixture underwent quick cooling to room temperature. In the second step, thermal annealing of the previously melted glass matrix was carried out at 560°C for 02 and 20 hours in order to enhance the diffusion of $Cd^{2+}$, $Mn^{2+}$, and $S^{2-}$ species into the host matrix. As a result of the thermal annealing, CdS and $Cd_{1-x}Mn_xS$ (x>0) NCs were formed in the glass template, wich were denominated at two classes: i) SNAB: CdS NCs and ii) SNAB: $Cd_{1-x}Mn_xS$ NCs [7].

### 2.3. Synthesis of $Zn_{1-x}Mn_xO$ Nanocrystals

Preparation of the $Zn_{1-x}Mn_xO$ NC samples is based on the transformation of the aqueous-$[Zn(NH_3)_4]^{2+}$ metal-complex in the presence of aqueous-$Mn^{2+}$, sodium oleate and hydrazine sulfate at 80°C. The best chemical synthesis results were achieved by keeping the pH value of the reaction medium at 8.5 during the whole reaction process, which was adjusted by controlling the addition of 4 M sodium hydroxide aqueous solution. Briefly, a typical protocol used started by magnetically-stirring, at room-temperature and for ½-hour, 100 mL of 0.38 M zinc chlorite mixed with 100 mL of 1.6 M ammonia hydroxide in order to form the aqueous-$[Zn(NH3)_4]^{2+}$ complex. Then 1 mL of hydrazine sulfate and 0.08 g of sodium oleate were added into the previously-stirred solution. The obtained reaction medium was then heated at 80°C using water-bath in order to transform the aqueous-$[Zn(NH3)_4]^{2+}$ complex, while keeping the pH value fixed at 8.5. The chemical synthesis was carried out for 2 h, while ammonia was observed to be release out from the reaction medium as the chemical process proceeded. The resulting precipitates ($Zn_{1-x}Mn_xO$; x≥0) were percolated and washed with distilled water and absolute ethanol for several times and further dried under at 500°C for 2h [5].

## 3. Optical properties of Nanocrystals

The optical properties of PbS, $Zn_{1-x}Mn_xO$ and $Cd_{1-x}Mn_xS$ NCs, synthesized by methodology describe in section 2, were investigated by Optical Absorption (OA), and/or Photoluminescence (PL) spectroscopy techniques. The obtained results will be presented and discussed as follows.

### 3.1. Optical properties of PbS Nanocrystals

#### *3.1.1. Optical Absorption and Photoluminescence*

Because of the large exciton Bohr radius (18 nm), PbS QD-doped glasses exhibit strong three-dimensional quantum-confinement effects at moderate nanocrystal size. This combined with small band gap energy (0.41 eV at room temperature) of PbS with different thermal annealing, which result in different average sizes of PbS nanocrystals [2,14]. Room-temperature Photoluminescence and optical absorption spectra of PbS nanocrystals with different time annealing process are shown in Figure 1. The strong quantum confinement in these structures is clearly observed. The appearance of defined band peaks in both absorption and emission spectra demonstrates the high quality of our samples and relatively small size distribution of the PbS nanocrystals.

#### *3.1.2. Size-dependent Stokes shift*

The dependence of the Stokes shift is closely linked to the size of the nanocrystals for schemes of strong quantum confinement the discrete levels of transition electron become more evident such that the difference between the position of peak absorption and emission increases with decreasing the size of the nanocrystals [14, 19]. With the absorption of a photon from of the valence band to the conduction band occurs the formation of electron-hole pairs (exciton). The exciton, once formed after absorption, cannot decay to the top of the valence band by a direct dipole transition and hence is denominated of dark exciton [20]. In the process of deexcitation eventually takes place with the help of phonons, thus giving rise to red shifted photons, known with Stokes shift.

The behavior of Stokes shift of PbS NCs of Figure 1 is represents in Figure 2. From the data presented is observed the decline in the Stokes shift with the increase of NCs size.

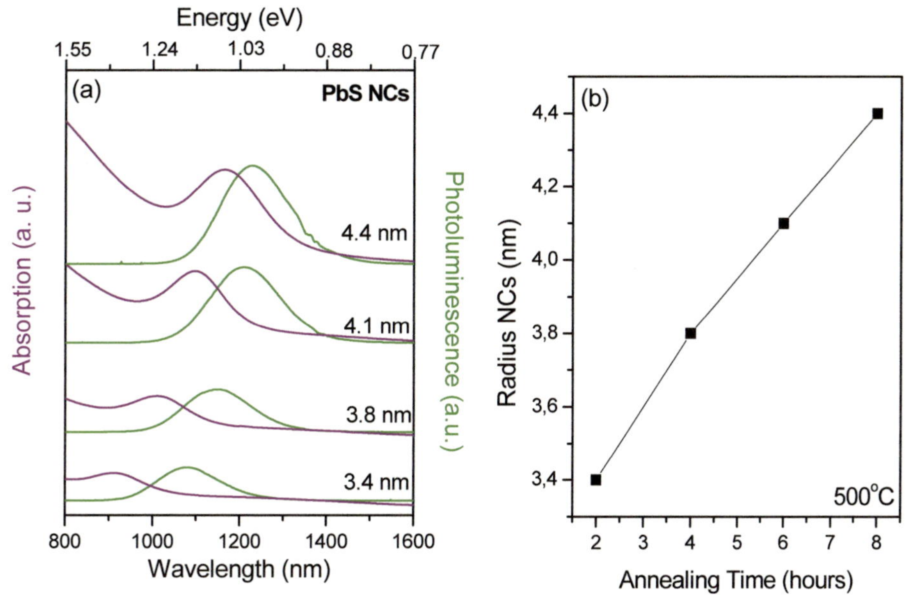


**Figure 1.** (a). Room-temperature optical absorption and photoluminescence of PbS NCs synthesized in SNABP glass matrix. (b) Behavior of average size of PbS NCs with annealing time at 500ºC.

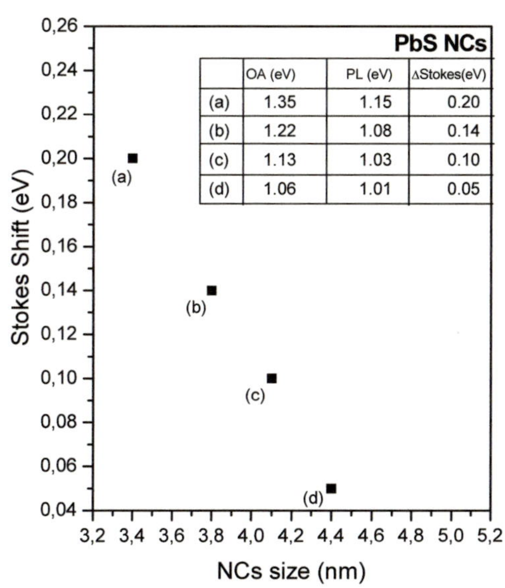


**Figure 2.** Behavior of Stokes shift to of PbS NCs synthesized in glass matrix.

The Stokes shift is originated by process of radiative decay from electron-hole recombination and a nonradiative decay via trapped states involves electron-phonons. As the size of the nanocrystal increase, the surface to volume ratio decreases, and there is a reduction in the overlap of the electron and hole wave functions. This is coincident with a decreased wave function overlap with the nanocrystal surface, which leads to less surface trapping and the decrease of Stokes shift [14]. A schematic of levels energy involved the process of excitation and deexicitation is shows in Figure 3. The excitation occurs with the absorption of electrons of level $1S_h$ (fundamental state) to the excitation states $1S_e$ ($OA_1$) and $1P_e$ ($OA_2$) and the deexcitation process is characterized by a nonradiative recombination of levels $1S_e$ and $1P_e$ to the surface trapped states and a radiative recombination (PL) to the level $1S_h$, clear observed in Figure 1(a) to PbS NCs.

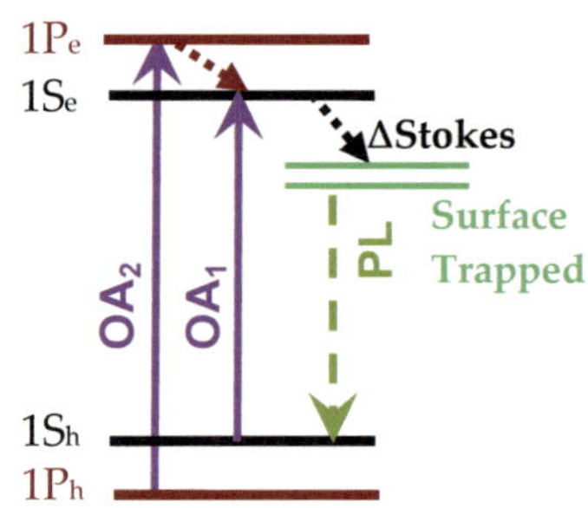


**Figure 3.** Model schematic used for explain the Stokes shift data difference between OA via radiative and nonradiative emission processes.

### 3.2. Optical properties of Zn1-xMnxO nanocrystals

Optical absorption spectra provide strong evidences of the $Mn^{2+}$ ions incorporation into the $Zn_{1-x}Mn_xO$ NCs different for the observed in samples with ZnO NCs. With the introduction of impurities magnetic in semiconductor NCs the optical properties are completely modify due the exchange interactions (*sp-d*) between electronic subsystem of NCs and electrons originated in the partially filled of the $Mn^{2+}$ ions. This exchange interactions causes the blue-shift of band gap observed in $Zn_{1-x}Mn_xO$ NCs in relationship to ZnO NCs that is proportional with the increase of x as show in Figure 4, for example, to x = 0 and 0.0081, being observed the blue-shift of band-gap of 3.33 eV (372 nm) to 3.41 eV (363 nm), respectively. This due the band gap of $Zn_{1-x}Mn_xO$ NCs semiconductor is between the ranges of 3.29 eV (gap ZnO bulk) at 4.2 eV (gap MnO bulk). The appearances of well-defined subband peaks in absorption spectra demonstrate the high quality of the synthesized samples and the relatively small size distribution of the NCs. Using this information with the energy of gap obtained by optical absorption spectra, the NCs size were determined by the effective mass model approximation of equation 01 [21]:

$$E \cong E_{bulk} + \frac{\hbar^2 \pi^2}{2eR^2}\left(\frac{1}{m_e m_o} + \frac{1}{m_h m_o}\right) \tag{1}$$

Where $E$ is band gap of the nanocrystals, $E_{bulk}$ is the band gap of the bulk material, $R$ is the particle radius, $m_e$ and $m_h$ are effective mass of the electrons and holes, respectively, and $m_o$ is the free electron mass. With the effective masses of electrons ($m_e = 0.28\ m_o$) and holes ($m_h = 0.59\ m_o$), we obtain the diameter of 4.1 nm for the as-prepared $Zn_{0.9919}Mn_{0.0081}$ NCs.

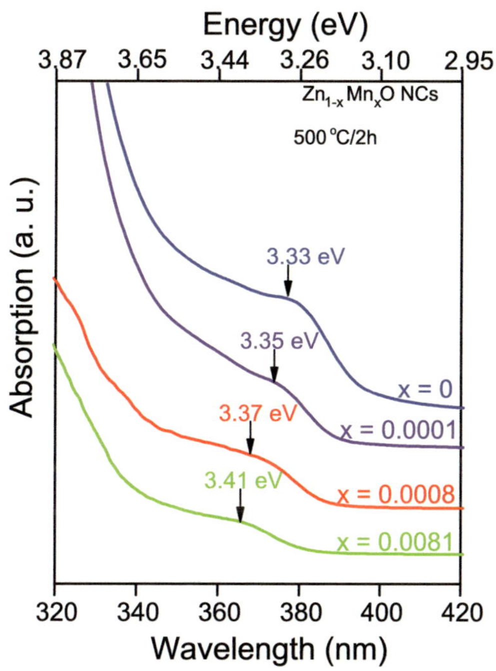


**Figure 4.** Room-temperature optical absorption spectra of $Zn_{1-x}Mn_xO$ NCs for x = 0, 0.001, 0.0008 and 0.0081.

### 3.3. Optical properties of Cd1-xMnxS nanocrystals

Figure 5 shows room-temperature OA spectra of the SNAB : $Cd_{1-x}Mn_xS$ set of samples with nominal concentrations x = 0, 0.001, 0.050, and 0.100. The quantum confinement can be clearly observed, and there are well-defined intersubband transition peaks in all OA spectra, which demonstrates the fairly good quality and narrow dot size distribution (~ 6%) in the samples. For a fixed doping concentration, an effective band gap reduction is observed, with the quantum confinement regime decreasing for increasing nanocrystal size, R. Analyzing these OA spectra and using a quantum confinement model based on effective-mass approximation, it is possible to estimate the average radius R of these dot samples using the expression [28] $E_{conf} = E_g + \hbar^2\ \pi^2/2\mu R^2 - 1.8(e^2/\varepsilon R)$, where $E_g$ is the energy band gap of the material (bulk), μ (ε) is the heavy-hole exciton reduced effective mass (dielectric constant), $e$ is the elementary charge, and the last term is the electron–hole effective Coulomb interaction.

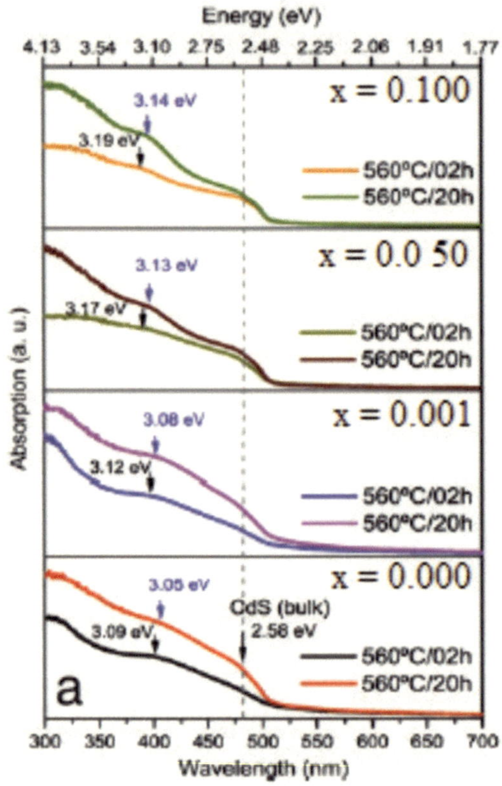


**Figure 5.** Panel (a): Room-temperature OA spectra of $Cd_{1-x}Mn_xS$ NCs embedded in SNAB glass matrix with concentration x = 0, 0.001, 0.050, and 0.100. The well-defined sub-band peaks demonstrate quantum confinement regimes and the relatively narrow dot size distributions. The energy shift of the OA peaks indicates the Mn incorporation in the dots.

The estimated average radii were R = 2.14 and R = 2.234 nm, for the samples annealed for 2 and 20 h, respectively. Under low magnetic impurity density and annealing temperature conditions, the dot size remained almost unchanged during Mn-ion incorporation induced by thermal treatment. In Fig. 3, we show the blue shift in the OA peaks, a quantity proportional to the concentration of magnetic ions. This shift changes from 3.09 eV in pure CdS (x = 0) dots to 3.19 eV in doped SNAB : $Cd_{0.900}Mn_{0.100}S$ samples treated at 560 °C for 2 h. For samples treated for 20 h, the values are 3.05 eV for x = 0 (undoped) and 3.14 eV for x = 0.100 (doped) dots.

## 4. Magnetic properties of DMS nanocrystals

The magnetic properties of DMS NCs are influenced by exchange interactions sp-d between the electronic subsystems of magnetic ions with the NCs, changing the configuration of confined electronic states. The $Zn_{1-x}Mn_xO$ and $Cd_{1-x}Mn_xS$ NCs were investigation by Electron Paramagnetic Resonance, $Cd_{1-x}Mn_xS$ and $Pb_{1-x}Mn_xS$ NCs investigation by Microscopy force Atomic (MFM) and $Pb_{1-x}Mn_xS$ NCs investigated by Magnetization measurements.

### 4.1. Evidencing the $Zn_{1-x}Mn_xO$ nanocrystal growth by Electron Paramagnetic Resonance

The EPR spectra of the $Zn_{1-x}Mn_xO$ NC samples with x = 0, 0.0001, 0.0008 and 0.0081 are shown in Figure 6(a). We found ZnO NCs sample (x = 0) presenting a sharp EPR signal with $g$ = 1.9568. However, the $Zn_{1-x}Mn_xO$ NC samples with low x-values exhibits a well-resolved EPR sextet in addition to some fine structure. They are stemmed from the hyperfine interaction between electron ($S$= 5/2) and nuclear ($I$ = 5/2) spins of the incorporated manganese ions, which are claimed to be located at different positions within the nanocrystal. For instance, the x = 0.0001 sample displays two sets of EPR sextets. Among them, the six well-defined EPR lines with hyperfine interaction splitting of 7.8 mT is assigned to isolated $Mn^{2+}$-ions substitutionaly incorporated into the $Zn_{1-x}Mn_xO$-core nanocrystal. While, the second set of EPR sextet structure, with hyperfine splitting of 8.0 mT, is due to $Mn^{2+}$-ions incorporated into the $Zn_{1-x}Mn_xO$-shell nanocrystal, at crystallographically-distorted sites near the NCs surface [27]. For this sample the smallest average diameter among the doped samples presents the strongest surface effect. Therefore, in comparison to the undistorted Mn-site (core-sites) the x = 0.0001 $Zn_{1-x}Mn_xO$ NCs sample holds enough distorted Mn-sites (shell-sites) to be probed by EPR in spite of low doping concentration. In addition, the sensitivity of the EPR technique to prove both core-like as well as shell-like Mn-ions with concentration as low as x = 0.0001 is a strong indication of the monodispersity in size. As the Mn-ion concentration increases, the number of EPR lines also increases, and the identified hyperfine structure is now superimposed to a broad EPR background line, as shown in EPR spectrum of the x = 0.0008 NCs sample. The underlying physics can be understood in the following way. As the Mn-ion concentration increases, the amount of Mn-ions in shell-sites increases. Hence, the replacement of Zn-ion by Mn-ions in various crystal sites with different distortion of crystal field occurs, resulting in a multi-line hyperfine structure. Furthermore, the increased concentration of manganese in the shell of nanocrystals may lead to the formation of Mn ion cluster, which inducing a strong Mn-Mn interaction. Hence broad EPR background line emerges. As the manganese concentration goes over x = 0.0081, however, the multi-line hyperfine structure collapses and the EPR spectra are replaced by a symmetric, broad single line due to enhanced Mn-Mn interaction. In order to confirm this analysis we have performed EPR spectral simulation of Mn-doped ZnO NC samples using time dependent perturbation theory [27], in which the spin-Hamiltonian is described by $\hat{H} = \hat{H}_z + \hat{H}_0$. In the spin-Hamiltonian $\hat{H}_z = \mu_e \hat{S} \cdot g_e \cdot \vec{B}$ is the Zeeman term, where $\mu_e$, $g_e$, and $\vec{B}$ are the Bohr magneton, the Lande factor and the applied magnetic field, respectively. The second term of the spin-Hamiltonian is $\hat{H}_0 = D\left[S_z^2 - S(S+1)/3\right] + E\left(S_x^2 - S_y^2\right) + A\hat{S} \cdot \hat{I}$, where the first two terms describe the zero-magnetic field fine-structure splitting due to spin-spin interaction of electrons, which is nonzero only in environments with symmetries lower than cubic. The third term ( $A\hat{S} \cdot \hat{I}$ ) is stemmed from the hyperfine interaction between electron and nuclear spins, leading to the observed six-line pattern. Since the interaction constants $A$, $D$ and $E$ strongly depend upon the local crystal field characteristics in which the $Mn^{2+}$-ion is located, the EPR spectrum varies when the local $Mn^{2+}$-ion crystal symmetry changes from $Zn_{1-x}Mn_xO$-core to $Zn_{1-x}Mn_xO$-shell. According to the pattern of the EPR spectra we found from our data, spectral simulation were performed by following a three step procedure, as illustrated in Figure 5 (a). Firstly, we computed the hyperfine structure. Second, a broad background resonance feature was simulated. Finally, we summed over these two spectra to end up with the EPR spectrum of the $Zn_{1-x}Mn_xO$ NCs sample with x = 0.0008. Figure 6 (b) displays the calculated spectra for x = 0.0001, 0.0008 and 0.0081. Excellent quantitative agreement between the simulated and the experimental spectra were achieved for instance in the case of the x = 0.0008 NCs (inset) sample using $g$ = 2.0033, $A$ = 7.8mT D = 6.1 mT, E = 0.5 mT, and 0.8 mT linewidth for the hyperfine structure simulation and $g$ = 2.0033 and 60 mT linewidth for the broad background resonance calculation. Therefore, we could conclude that the EPR hyperfine six-lines are due to $\Delta m_S = \pm 1$ and $\Delta m_I = 0$ transitions, where $m_S$ ($m_I$) stands for the projection of the spin $S$ ($I$), the broad background is originated from the exchange narrowing due to the strong Mn-Mn interaction. Hence, the EPR simulations strongly support the picture that Mn-ions are incorporated into the hosting ZnO nanocrystals.

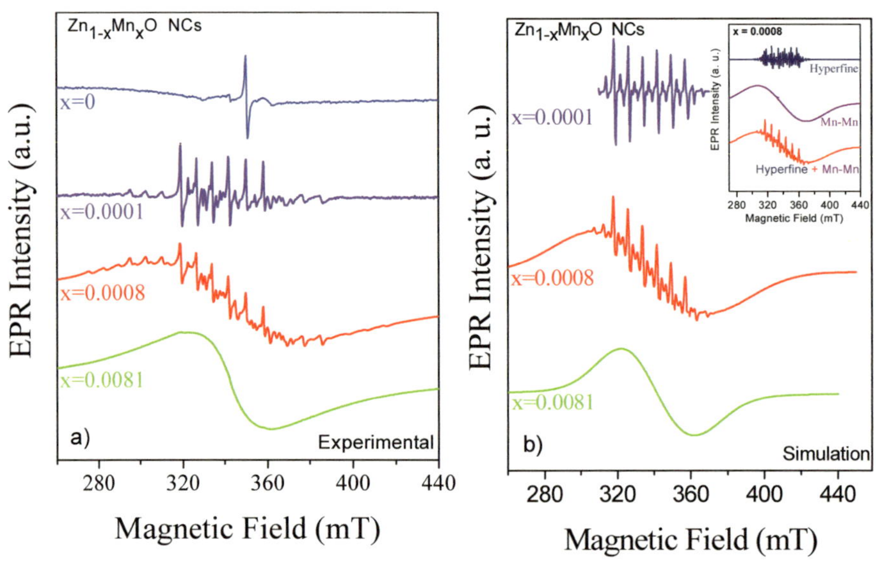


**Figure 6.** (a). Room-temperature X-band EPR spectra of $Zn_{1-x}Mn_xO$ nanocrystals at different x values and (b) Simulated EPR spectra of $Zn_{1-x}Mn_xO$ nanocrystals for different x values. In inset illustration of processes used for EPR spectra simulation of the $Zn_{1-x}Mn_xO$ nanocrystal (x = 0.0008).

## 4.2. Confirming the migration process of $Mn^{2+}$ ions in $Cd_{1-x}Mn_xS$ nanocrystals by Electron Paramagnetic Resonance

Theoretical model for explain the incorporation of magnetic impurities in nanocrystals are related in the literature [22-24], such as the "self-purification" mechanisms that are explained through energetic arguments. These mechanisms show that the formation energy of magnectic impurities increases when the NCs size decreases. Moreover, the binding energy of the impurities in the crystalline faces is highly dependent on the semiconductor material, such as the crystal structure and NCs shape [23].

The energy required to replace a $Cd^{2+}$ ion by an $Mn^{2+}$ ion, called the formation energy, is greater for smaller $Cd_{1-x}Mn_xS$ NCs [22]. Thus, the present $Mn^{2+}$ ions in smaller $Cd_{1-x}Mn_xS$ NCs are less stable, promoting the difusion of some of these impurities to regions closer to the dot surface, i.e., to the site $S_{II}$. This ability of the $Mn^{2+}$ ion to diffuse through the nanocrystal is quite reasonable because the ionic radius of impurity (83 pm) is smaller than the ionic radius of the $Cd^{2+}$ ion (95 pm). This mechanism, known as "self-purification" [24], is an intrinsic property of impurities (or defects) in semiconductor related to NCs size-dependent energetic arguments and explains the predominance of the signal $S_{II}$ on the $S_I$ in the EPR spectra shown in the Figure 7.

The EPR spectra shown in Figure 7(a) for selected SNAB : $Cd_{1-x}Mn_xS$ samples with x = 0.001, 0.050, and 0.100 display well-resolved transitions between a sixplet sublevel structure inserted into a broader horizontal S-like shaped EPR structure typical for free-like electron states with spin S = 1/2, ±1/2. These six sublevels are associated with the magnetic quantum numbers $M_S$ = ±5/2 ,± 3/2, ± 1/2 that occur in fine exchange interaction induced transitions when Mn ions are present in a sample. Another much weaker interacting sixplet sublevel structure occurs because the hyperfine interaction coupling between the spin of localized electrons ($S$ = $1/2$) with nuclear spin ($I$ = $5/2$) of the incorporated Mn-ions in the doped samples. Among them, are six fairly welldefined EPR lines with fine interaction splitting described by the simulation signal SI, and assigned to dilute concentration of $Mn^{2+}$ ions found in substitutional Cd places inside the CdS nanocrystal cell. However, the second EPR sixplet structure set described by the simulation signal SII is due to Mn ions located at crystallographically highly distorted sites near the dot surface.

In order to confirm this analysis we have performed EPR spectral simulation of Mn-doped CdS samples using time-dependent perturbation theory in which the spin Hamiltonian is given by $H = H_Z + H_0$, where $H_Z = \mu_e \vec{S} \cdot g_e \cdot \vec{B}$ describes the Zeeman interaction and $\mu_e$, $g_e$, and $\vec{B}$ represent the Bohr magneton, the Landé g-factor, and the applied magnetic field, respectively. The second term, $H_0 = D\left[S_z^2 - S(S+1)/3\right] + E\left(S_x^2 - S_y^2\right) + A\hat{S}\cdot\hat{I}$, includes the zero magnetic field fine interaction (terms proportional to coupling constants D and E) between the electron spin and the crystal field.

This contribution only induces non-zero fine structure splitting in crystalline environments with symmetries lower than cubic. Finally, the term $A\hat{S}\cdot\hat{I}$ represents the hyperfine interaction between localized electrons and nuclear spins in the Mn ions and where each electron transition splits into six additional levels characterized by the nuclear magnetic quantum numbers ($M_I$ = ±5/2,±3/2,±1/2), producing, in principle, a total of 36 transitions. However, selection rules limit the number of allowed transitions, and the broader and stronger features observed in the EPR spectra are due to lines associated to the dipole-allowed $\Delta M_S$ = ±1 transitions with $\Delta M_I$ = 0. Since the interaction constants A, D, and E

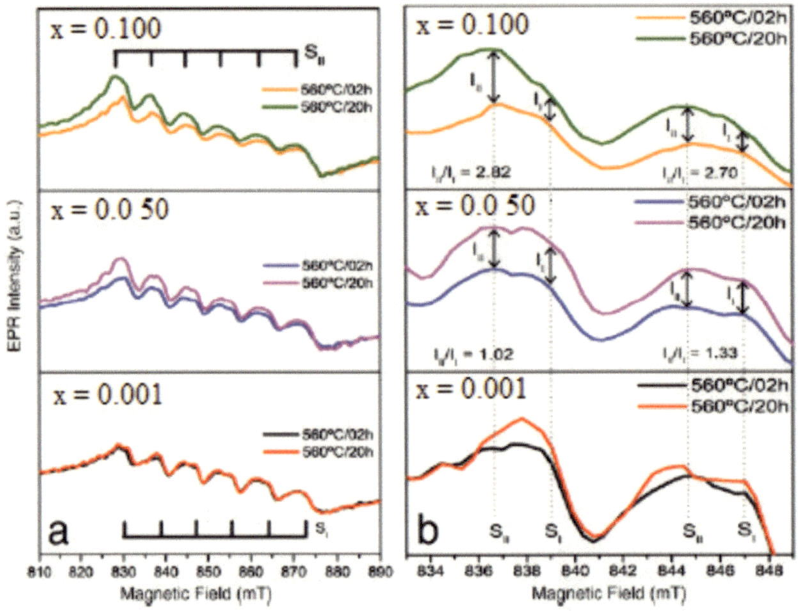


**Figure 7.** Panel (a): Room-temperature EPR spectra of selected $Cd_{1-x}Mn_xS$ NCs embedded in the SNAB glass matrix with Mn concentration x = 0.001, 0.050, 0.100. Panel (b) shows a zoom on two of the sublevel peaks, between 835 and 850 mT, in order to observe the intensity differences, marked as $I_{II}$ and $I_I$. Noted the change of the ratio between EPR intensities when Mn ions are placed at the dot interface ($I_{II}$) or inside the core cell ($I_I$), in samples with concentrations x = 0.050 and x = 0.100. This ratio change indicates the interdiffusion of magnetic ions due to thermal treatment.

strongly depend on the local crystal field characteristics near the $Mn^{2+}$ ion location, the EPR spectrum will be changed when substitutional $Mn^{2+}$ ions move from the $Cd_{1-x}Mn_xS$ dot core (signal $S_I$) to the dot surface (signal $S_{II}$) region. On the other hand, any isolated Mn ion that is dispersed inside the glass matrix cannot be identified in the EPR spectra due to the absence of a well-defined crystal field of this amorphous material (glass). In addition, the Mn–Mn interactions would also be intensified if the formation of Mn clusters occurred in the glass environment. However, we consider that this later effect may be neglected, since the formation of Mn clusters is highly unlike due to the small amount of incorporated Mn ions in these samples.

The other enhancement on the EPR signal intensity, as observed in Figure 7(a), can be associated to the increased concentration of $Mn^{2+}$ ions. The larger the Mn density is, in samples subjected to thermal treatment for 20 h and also seen in the AFM images, the stronger is the EPR intensity. In addition, the EPR intensity can be strengthened due to migration of a fraction of magnetic ions from the glass matrix to the surface of the NCs, an effect which also increases the intensity of the broader background EPR peak due to Mn–Mn interaction.

The zoom in two of the sublevel exchange peaks, between 835 and 850 mT, displayed in Figure 7(b) shows the enhancement of all EPR lines for longer annealing times. However, the EPR intensity for surface Mn ions $I_{II}$ increases whereas for the substitutional ions $I_I$ decreases in both samples with nominal concentrations x = 0.050 and x = 0.100, when the thermal treatment increases from 2 to 20 h. The change in the ratio between intensities, $I_{II} / I_I$, confirms the migration of a fraction of Mn ions incorporated in the dot core to the surface region.

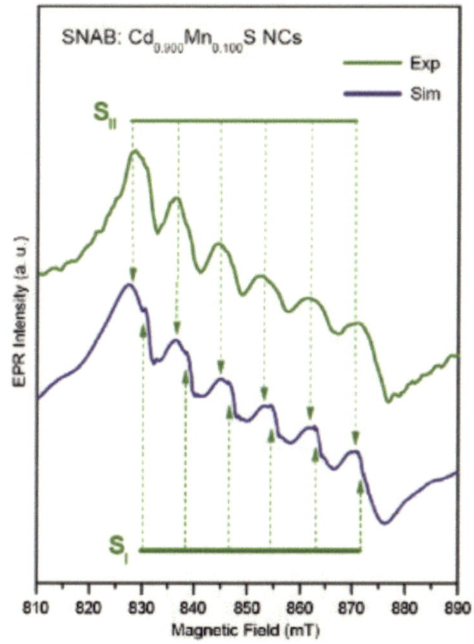


**Figure 8.** Room-temperature EPR spectrum in $Cd_{0.900}Mn_{0.100}S$ NCs measured in the K band. The simulated EPR transitions were obtained by the combination of a stronger spectrum associated to Mn replacing the Cd ions inside the wurtzite cell (signal $S_I$) and a weaker spectrum due to ions located at or near the dot surface region (signal $S_{II}$). The Hamiltonian model describing the fine and hyperfine magnetic interactions contributing to dipole transitions between the eigenvalues is discussed in the text.

The energy required to replace $Cd^{2+}$ by $Mn^{2+}$ in the crystal, usually referred to as the ''formation energy'', is larger for small- size CdS NCs [9]. Therefore, $Mn^{2+}$ ions in the less stable places inside $Cd_{1-x}Mn_xS$ dots are moved to interface region in a process of energy minimization that promotes the interdiffusion of a fraction of impurities from the core (signal $S_I$) to larger binding energy regions near [24] to the surface and, in this location, the magnetic ions generate the signal $S_{II}$. The ability of $Mn^{2+}$ ions to migrate through the NC is quite reasonable because the ionic radius of this magnetic impurity (83 pm) is smaller than the ionic radius of the $Cd^{2+}$ ion (95 pm). This mechanism seems to be an intrinsic general property of impurities (or defects) in semiconductor NCs [24], and the shape and the crystal structure of a NC determine which surface is more favorable for impurity binding; these facets are the (0001) crystalline planes in the CdS wurtzite structure [9]. Here, the larger ratio III/II between intensities of EPR peaks in samples with different annealing times, as shown in the Figure 7(b), corroborates this idea of controlled migration process.

The simulated signals for the EPR transitions in the $Cd_{0.900}Mn_{0.100}$ S sample, shown in Figure 8, confirm the presence of $Mn^{2+}$ ions in two distinct sites: incorporation occurring in the core (signal $S_I$), and near the surface (signal $S_{II}$). The hyperfine interaction constants used to simulate these spectra were $A_I = 8.1$ mT and $A_{II} = 8.4$ mT for a magnetic system with spin S = 5/2, nuclear spin I = 5/2, and g-factor $g_e = 2.005$. The fine structure constants were D = 40 mT and E = 5 mT. We believe that these results have shown unambiguously that proper use of annealing temperature with different times may produce controlled diffusion of $Mn^{2+}$ ions in magnetic dots. These findings are strongly supported by the fairly good agreement between simulated and experimental EPR resonant transitions.

### 4.3. Investigating the $Pb_{1-x}Mn_xS$ nanocrystals by susceptibility and magnetization measurements

Figure 9 shows the temperature dependence of the inverse of the real part of the magnetic susceptibility ($1/\chi$) of $Pb_{1-x}Mn_xS$ NCs growth in borosilicate glass with annealing at 500°C at 10hours, for x = 0.003 and 0.007, recorded at 100 Oe. The susceptibility data presented in Figure 9 follow the Curie-Weiss law, $\chi(T) = C/(T-\theta)$, where $\theta$ is the Curie-Weiss temperature [25, 26]. The fitted value found for the parameter $C$ in $\chi(T)$ allowed estimation of the isolated $Mn^{2+}$ ion content incorporated in the $Pb_{1-x}Mn_xS$ NCs embedded in glass matrix using:

$$x = (m_A + m_B)/\left[S(S+1)(g\mu_B)2N_A/(3k_BC) + m_A - m_M\right] \quad (2)$$

where $m_A$, $m_B$ and $m_M$ represent atomic mass for the cation (Pb), anion (S) and magnetically doped (Mn), respectively. $S$ represents the total spin of the $Mn^{2+}$ ion, $k_B$ is the Boltzmann constant, and $N_A$ is Avogadro's number. In this calculation $g = 2$ is assumed for the $Mn^{2+}$ ion.

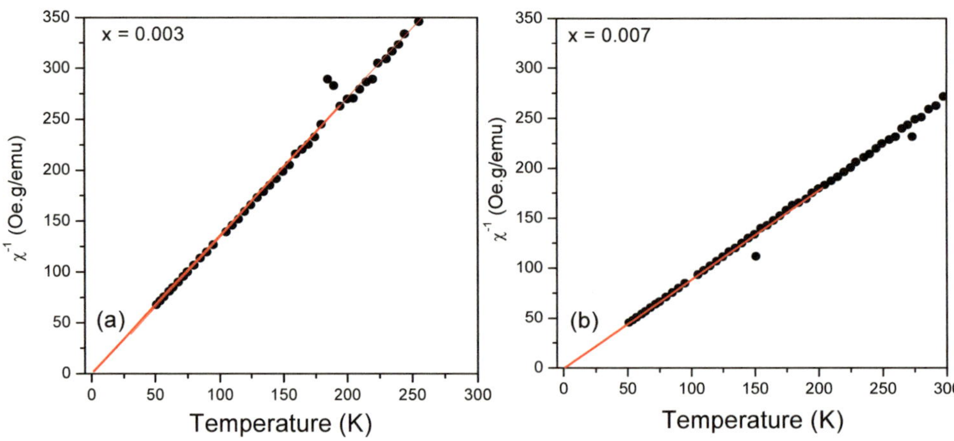


**Figure 9.** The temperature dependence on the inverse of susceptibility, recorded at 100 Oe, shows low temperature paramagnetic character for $Pb_{1-x}Mn_xS$ NCs embedded in borosilicate glass matrix.

Using the equation 02, estimates the concentration of Mn ions incorporated into the crystal lattice of the nanocrystals. For the nominal concentrations of x = 0.003 and x = 0.007 were estimated real concentrations of x = 0.003 and x = 0006 respectively, resulting in the formation of $Pb_{0.997}Mn_{0.003}S$ and $Pb_{0.994}Mn_{0.006}S$ NCs. Field-cooled magnetization curves of $Pb_{0.997}Mn_{0.003}S$ and $Pb_{0.994}Mn_{0.006}S$ NCs growth in borosilicate glass matrix, recorded with an external applied field of 1 KOe and over a wide range of temperatures, are shown in Figures 9(a) and (b), respectively. Each inset in these figures displays the T=1.28 K field dependence of the magnetization.

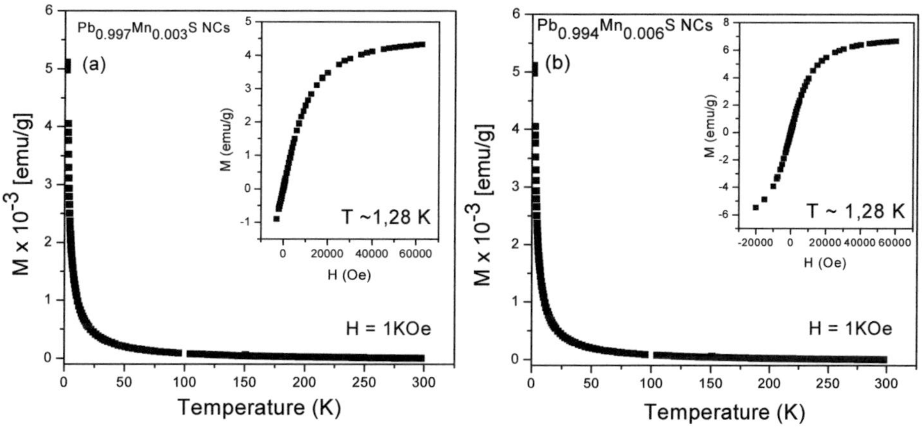


**Figure 10.** Magnetization as a function of temperature for the $Pb_{0.997}Mn_{0.003}S$ and $Pb_{0.994}Mn_{0.006}S$ NCs embedded in borosilicate glass matrix. Field dependence of magnetization for each sample is shown in the insets.

The magnetic data presented in Figure 9 including the corresponding insets, reveal the dominant paramagnetic behavior of the dots.

## 4.4. Magnetic force microscopy of semimagnetic nanoparticles: $Pb_{1-x}Mn_xS$ nanocrystals

Figure 10 presents the AFM/MFM images for the $Pb_{0.993}Mn_{0.007}S$ NCs samples subjected to the thermal annealing for 10 h, where we confirmed the high density of the nanocrystals with quantum confinement properties as well as bulk-like properties. In figure 10 (a), the AFM image is also influenced by the sample topography, while in figure 10(b), in the MFM images there are only the magnetic interactions, confirming the formation of semimagnetic $Pb_{0.993}Mn_{0.007}S$ NCs, with average NCs-size of approximately 6.0 nm. In an attractive configuration, the NCs have magnetization in a parallel direction to the tip magnetization, resulting in dark areas of the MFM image. However, in a repulsive configuration, the NCs have magnetization in an antiparallel direction to the tip magnetization, resulting in bright areas of the MFM image. Therefore, the formation of the dark and bright fields related to a single spin domain in the $Pb_{0.993}Mn_{0.007}S$ NCs is shown by the clear contrast in these MFM images.

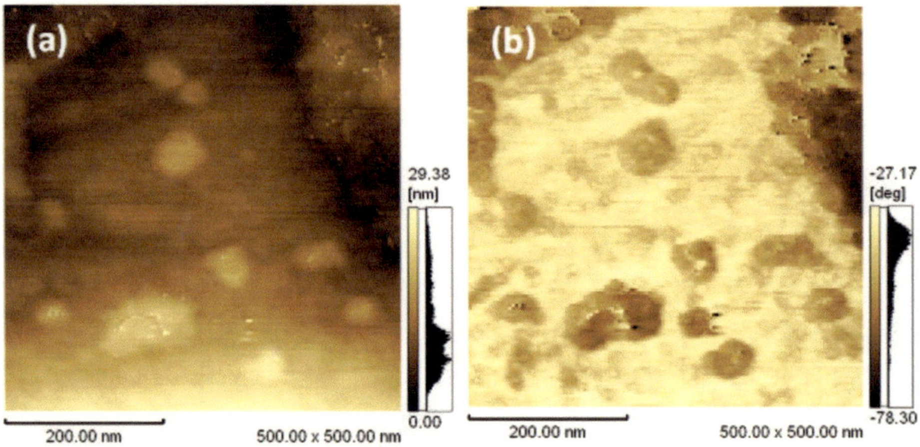


**Figure 11.** AFM/MFM of $Pb_{0.993}Mn_{0.007}S$ NCs growth in borosilicate glass matrix.

## 5. Structural properties of $Zn_{1-x}Mn_xO$ nanocrystals

The structural properties of $Zn_{1-x}Mn_xO$ nanocrystals were investigated by X-ray diffraction and Raman spectroscopy techniques.

### 5.1. Characterizing of $Zn_{1-x}Mn_xO$ nanocrystals by X-ray diffraction

The XRD patterns of the $Zn_{1-x}Mn_xO$ ($x \geq 0$) NCs samples for x = 0, x = 0.0001, x = 0.0008 and x = 0.0081 are shown in Figure 11. It is noted that the typical bulk-ZnO hexagonal wurtzite crystal structure is preserved in the as-precipitated $Zn_{1-x}Mn_xO$ ($x \geq 0$) NCs samples whit treatment thermal of 500°C by 2 hours [27]. Except for the ZnO with treatment thermal the 60°C by 12 hours that is present in amorphous phase. Nevertheless, the characteristic XRD peaks shift towards lower diffraction angle values as the Mn-ion concentration in the hosting ZnO structure increases, as shown in inset for (002) peak, clearly indicating an increase of the lattice constant. Using the Cohen's method we performed the estimation of the c-axis lattice constant from the following three selected XRD peaks: (100), (002), and (101). One found that the average c-axis lattice crystal constant of the x = 0, 0.0001, 0.0008 and 0.0081 $Zn_{1-x}Mn_xO$ NC samples were 5.207, 5.208, 5.226 and 5.231 Å, respectively. The monotonic increase observed in the lattice crystal constant is attributed to the replacement of the $Zn^{2+}$-ion, with smaller ionic radius (0.74 Å in the hexagonal wurtzite ZnO crystal structure), by $Mn^{2+}$-ion with larger ionic radius (0.83 Å). These XRD findings provide evidences that the as-produced samples are high-quality and single-phased $Zn_{1-x}Mn_xO$ crystals in the nanoscale regime.

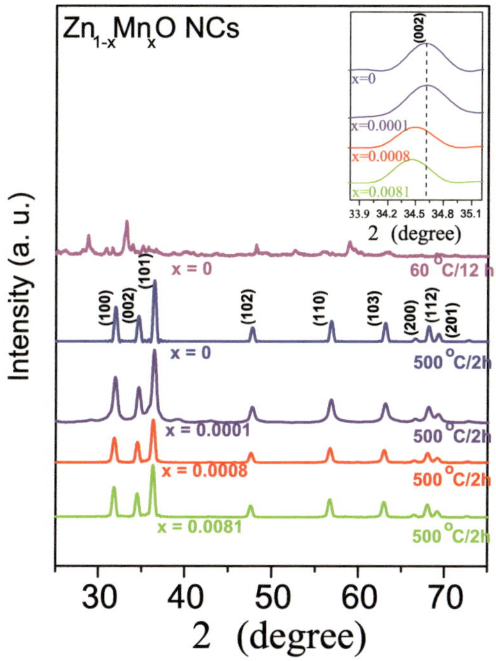


**Figure 12.** X-ray diffraction of ZnO bulk and $Zn_{1-x}Mn_xO$ nanocrystal samples with x = 0, 0.0008 and 0.0081. In inset shift of the (002) $Zn_{1-x}Mn_xO$ nanocrystal XRD peak as a function of Mn-ion concentration (x).

## 5.2. Characterizing of $Zn_{1-x}Mn_xO$ nanocrystals by Raman spectroscopy

Figure 12 shows the Raman spectra of ZnO bulk and $Zn_{1-x}Mn_xO$ NCs for x = 0, 0.0008 and 0.0081. The wurtzite ZnO structure belongs to the space group $C_{6v}^4$ with two formula units per primitive cell. Therefore, group theory predicts that the zone-center optical phonons are described by $\Gamma_{opt} = A_1 + 2B_1 + E_1 + 2E_2$ [5]. The $A_1$ and $E_1$ modes represent Raman and infrared active polar phonons, showing frequencies for transverse-optical (TO) and longitudinal-optical (LO) modes. The $E_2$ mode is non-polar and is Raman active in two frequencies; the $E_2$ (high) associated to the oxygen anions and the $E_2$ (low) associated to the Zn cations in the lattice. Finally, the $B_1$ mode is Raman inactive. The Raman peak centered at 334 $cm^{-1}$ is described through a multi-phonon process associated to three different modes; the dominant $A_1$ mode plus a weak $E_2$ component and an even weaker $E_1$ component. The literature describes the frequency of this mode as the difference between the $E_2$ (high) and $E_2$ (low) modes. The $A_1$ (TO), $E_1$ (TO), $A_1$ (LO) and $E_1$ (LO) modes were observed at 382, 410, 541 and 586 $cm^{-1}$, respectively. The Raman peak observed at 439 $cm^{-1}$ represents the $E_2$ (high) mode associated to the oxygen anions. The Raman peaks of $Zn_{1-x}Mn_xO$ NCs show a shift to low frequencies and the peaks are asymmetry in relation to the ZnO semiconductor. With the incorporation of $Mn^{2+}$ ions is observed a peak was observed at about 659 $cm^{-1}$ becoming more intense with increasing of Mn-concentration in $Zn_{1-x}Mn_xO$ NCs. The 659 $cm^{-1}$ peak is associated the two additional modes $\left[A_1(LO) + E_2(low)\right]$ originated from the precipitation phase of $ZnMn_2O_4$ as observed for x = 0.0008 and 0.0081 in the samples of $Zn_{1-x}Mn_xO$ NCs. This finding is a strong support to the picture that Zn-ions in the ZnO crystal structure are replaced by Mn-ions during the course of the chemical precipitation process.

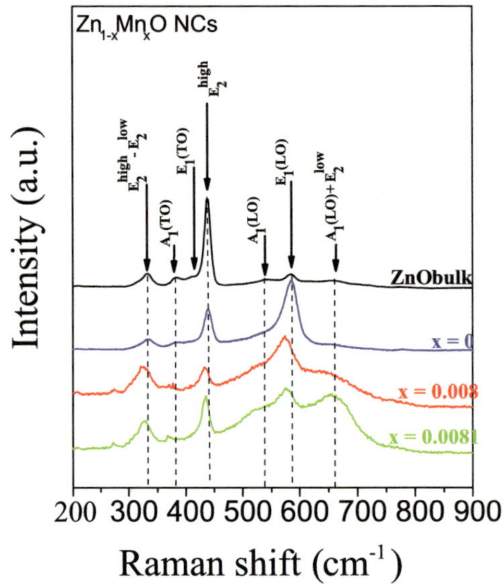


**Figure 13.** Room-temperature Raman spectra of ZnO bulk and $Zn_{1-x}Mn_xO$ NCs for x = 0, 0.0008 and 0.0081.

## 6. Conclusions

In conclusions, we report the successfully synthesis of semiconductors and semimagnetic nanocrystals by different methodologies.

PbS, $Pb_{1-x}Mn_xS$ and $Cd_{1-x}Mn_xS$ NCs were growth in borosilicate glass by fusion method and $Zn_{1-x}Mn_xO$ NCs were synthesized by co-precipitation method.

The investigation of semiconductor and semimagnetic NCs provided by experimental techniques of Optical Absorption, Photoluminescence, Electron Paramagnetic Resonance, Magnetic Force Microscopy, X-Ray Diffraction, and Raman spectroscopy, have revealed the control of optical, magnetic and structural properties and the high-quality of NCs synthesized by different methodologies.

We believe that this chapter can motivate inspire further investigation of these systems in a search for possible device applications.

## Author details

Ricardo Souza da Silva
*Instituto de Ciências Exatas e Naturais e Educação (ICENE),*
*Departamento de Física, Universidade Federal do Triângulo Mineiro, Uberaba, Minas Gerais, Brazil*

Ernesto Soares de Freitas Neto and Noelio Oliveira Dantas
*Laboratório de Novos Materiais Isolantes e Semicondutores (LNMIS),*
*Instituto de Física, Universidade Federal de Uberlândia, Uberlândia, Minas Gerais, Brazil*

## Acknowledgement

The authors gratefully acknowledge the financial support from the Brazilian agencies: MCT/CNPq, Capes, Fapemig and FUNEPU. We are also grateful to our collaborators: Augusto Miguel Alcalde ( A. M. Alcalde), Eliane da Costa Vilela (E. C. Vilela), Felipe Chen Abrego (F. Chen), Fernando Pelegrini (F. Pelegrini), Gilmar Eugênio Marques (G. E. Marques), Henry Socrates Lavalle Sullasi (H. S. L. Sullasi), Jales Franco Ribeiro da Cunha (J. F. R. Cunha), Leonardo damigo (L. Damigo), Kely Lopes Caiado Miranda (K. L. Miranda), Marcelo de Assumpção Pereira da Silva (M. A. Pereira-da-Silva), Miguel Alexandre Novak (M. A. Novak), Patrícia Pommé Confessori Sartoratto (P. P. C. Sartoratto), Paulo César de Morais (P. C. Morais), Qu Fanyao (Qu Fanyao) and Victor Lopez Richard (V. Lopez-Richard).

## 7. References

[1] Norris D. J, Efros A. L, Erwin S. C (2008) Doped Nanocrystals. Science 319: 1776-1779.
[2] Dantas N. O, Qu Fanyao. Silva R.S, Morais P. C (2002) Anti-Stokes Photoluminescence in Quantum Dots. Jour. Of Phys. Chem. B. 106: 7453-7457.
[3] Silva R. S, Morais P. C, Qu fanyai, Alcalde A. M, Dantas N. O, Sullasi H. S (2007) Synthesis process controlled magnetic properties of Pb1-xMnxS nanocrystals. App. Phy. Lett.. 90: 253114-1-253114-3.
[4] Neto E. S. F, Dantas N. O, Neto N. M. B, Guedes I, Chen F (2011) Control of luminescence emitted by Cd1-xMnxS nanocrystals in a glass matrix: concentration and thermal annealing. Nanotechnology 22: 105709.
[5] Dantas N. O, Damigo L, Qu Fanyao, Qu Fanyao, Cunha J. F. R, Silva R. S, Miranda K. L, Vilela E. C, Sartoratto P. P. C, Morais P. C (2008) Raman investigation of ZnO and Zn1-xMnxO nanocrystals synthesized by precipitation method. J. Non-Cryst. Solids 354: 4827-4829.
[6] Gaponenko, S. V (1998). Optical Properties of Semiconductor Nanocrystals, Cambridge University Press. 260 p.
[7] Dantas N. O, Neto E. S. F, Silva R.S (2010) Diluted Magnetic semiconductors in Glass matrix. In: Masuda Y. Nanocrystals. Sciyo: InTech. pp. 143-168.
[8] Gur I, Fromer N. A, Geier M. L., Alivisatos A. P (2005) Air-stable all-inorganic nanocrystal solar cells processed from solution. Science 310: 462-465.
[9] Erwin S. C, Zu L, Haftel M. I,. Efros A. L, Kennedy T. ., Norris D. J (2005) Doping semiconductor nanocrystals. Nature: 436, 91-94.
[10] Timmerman D,Valenta J, Dohnalová K, de Boerl, W. D. A. M, Gregorkiewicz T (2011) Step-like enhancement of luminescence quantum yield of silicon nanocrystals. Nature Nanotechnology 6: 710-713.Volume:
[11] Vach H (2011) Ultrastable Silicon Nanocrystals due to Electron Delocalization. Nano Lett. 11: 5477–5481.
[12] Furdyna, J. K. (1988). Diluted magnetic semiconductors. J. Appl. Phys. 64: R29 –R64.

[13] Yu J. H, Liu X, Kweon K. E, Joo J, Park J, Ko K. –T, Lee D. W, Shen S, Tivakornsasithorn K, Son J. S, Park J. –H, Kim Y. –W, Hwang G. S, Dobrowolska M, Furdyna J. K, Hyeon T (2010) Giant Zeeman splitting in nucleation-controlled doped CdSe:Mn2+ quantum nanoribbons. Nature Materials 9: 47-53.

[14] Dantas N. O, de Paula P. M. N, Silva R. S, López-Richard V, Marques G. E (2011) Radiative versus nonradiative optical processes in PbS nanocrystals. J. Appl. Phys. 109: 024308-1 - 024308-4.

[15] Dantas N. O, Neto E.S.F, Silva R. S, Chen F, Pereira-da-Silva M. A, López-Richard V, Marques G. E (2012) The migration of Mn2+ ions in Cd1-xMnxS nanocrystals: thermal annealing control. Solid State Communications 5: 337-340.

[16] Silva R.S, Morais P. C, Mosiniewicz-Szablewska E, Cuevas R. F, Campoy J. C. P, Pelegrini F, Qu Fanyao, Dantas N. O (2008) Synthesis and Magnetic Characterization of Pb1-xMnxS Nanocrystals in Glass Matrix. J. Phys. D: Appl. Phys. 41: 165005-1 - 165005-5.

[17] Dantas N. O, Pelegrini F,Novak M. A, Morais P.C, Marques G. E, Silva R.S (2012) Control of magnetic behavior by Pb1-xMnxS nanocrystals in a glass matrix. J. of Appl. Phys 111: 106206-1 – 106206-5.

[18] Dantas N. O, Damigo L, Qu Fanyao, Silva R. S, Sartoratto P. P. C, Miranda K. L, Vilela E. C, Pelegrini F, Morais P. C. (2008) Structural and magnetic properties of ZnO and Zn1-xMnxO nanocrystals J. Non-Cryst. Solids 354: 4727 – 4729.

[19] Zhang J, Jiang X (2008) Confinement-Dependent Below-Gap State in PbS Quantum Dot Films Probed by Continuous-Wave Photoinduced Absorption. J. Phys. Chem. B 112: 9557-9560.

[20] Bagga A, Chattopadhyay P K, Ghosh S (2006) Origin of Stokes shift in InAs and CdSe quantum dots: Exchange splitting of excitonic states. Phys. Rev. B 74: 035341-1 - 035341-7.

[21] Brus, L. E (1983) A Simple-Model For the Ionization-Potential, Electron-Affinity, and Aqueous Redox Potentials of Small Semiconductor Crystallites. J. of Chem. Phys. 79: 5566-5571.

[22] Dantas N. O, Neto E.S.F, Silva R. S, Chen F, Pereira-da-Silva M. A, López-Richard V, Marques G. E (2012) The migration of Mn2+ ions in Cd1-xMnxS nanocrystals: thermal annealing control. Solid State Communications 5: 337-340.

[23] Norris D. J, Efros A. L, Erwin, S. C (2008) Doped Nanocrystals. Science 319: 1776-1779.

[24] Dalpian G. M, Chelikowsky J. R (2006) Self-Purification in Semiconductor Nanocrystals. Phys. Rev. Lett. 96, 226802-1 - 226802-4.

[25] Dantas N. O, Pelegrini F,Novak M. A, Morais P.C, Marques G. E, Silva R.S (2012) Control of magnetic behavior by Pb1-xMnxS nanocrystals in a glass matrix. J. of Appl. Phys 111: 106206-1 – 106206-5.

[26] Górska, M, Anderson, J. R (1988) Magnetic susceptibility and exchange in IV-VI compound diluted magnetic semiconductors. Phys. Rev. B 38: 9120-9126.

[27] Dantas N. O, Damigo L, Qu Fanyao, Silva R. S, Sartoratto P. P. C, Miranda K. L, Vilela E. C, Pelegrini F, Morais P. C. (2008) Structural and magnetic properties of ZnO and Zn1-xMnxO nanocrystals. J. Non-Cryst. Solids 354: 4727 – 4729.

[28] Brus L. E. (1984) Electron–electron and electronhole interactions in small semiconductor crystallites: The size dependence of the lowest excited electronic state. J. of Chem. Phys. 80: 4403-4409.

# Chapter 5

# Optical Nanocomposites Based on High Nanoparticles Concentration and Its Holographic Application

Igor Yu. Denisyuk, Julia A. Burunkova,
Sandor Kokenyesi, Vera G. Bulgakova and Mari Iv. Fokina

Additional information is available at the end of the chapter

http://dx.doi.org/10.5772/47737

## 1. Introduction

The main problem of current photolithography is diminishing of minimal feature sizes up to subwavelength value. The smallest feature size $X_{min}$ that can be projected by a coherent imaging system is $X_{min}=\lambda/2NA$, where $\lambda$ is the wavelength of the illumination and NA is the numerical aperture of the lens. The most ordinary way to attain smaller feature sizes is to reduce the wavelength up to 193 nm. The NA is typically 0,8, so the feature size is on the order of the exposure wavelength. State of the art semiconductor fabrication facilities in the year 2010 are forecasted to use a 32 nm process, in these conditions to achieve the required resolution and depth of field optical lithography techniques become increasingly difficult.

Now was developed new lithography technique - deep lithography that suitable for making elements with high aspect ratio. Here for vertical borders preparation it is need to use NA about 0,05 or less that limit resolution by value around 4 um.

As the fundamental limits of optical lithography are approached, the nonlinear, self-focusing and self-writing properties of the photoresist become increasingly important.

Special UV-curable nanocomposite with strong non-linear and self-writing effects used as a photoresist to improve light distribution in the spot as well in volume of photoresist can be used. Same proposed technique is applicable for deep lithography based on 365 nm UV light with high scattering to improve shape of small feature in results violation of the laws of geometrical optics at use light self-focusing in materials. If to make nanocomposite system with self-writing effects and place it as a topcoat, we will obtain self-writing subwavelength artificial waveguide that will guide the light to small subwavelength spot. So to make it it's necessary to develop special material with self-writing effects.

Same material having self-writing effects will be applicable for holographic lithography, i.e. writing of 3D lattice by interference of two, three or four coherent laser beams. Self-writing effects will increase modulation of refractive index of material and therefore diffraction efficiency of ready element.

Current work describes development of nanocomposite material, investigation of its holographic proprieties and sub wavelength lithography application based on self-writing effects in material.

## 2. Nanocomposite based on nanoparticles in UV-curable monomers mixture

### 2.1. Materials and methods

Chemicals: Monomers 2-Carboxyethyl acrylate (2Carb, Aldrich № 552348), Bisphenol A glycerolate (BisA, Aldrich № 41,116-7), 2-Phenoxyethyl acrylate (PEA,Aldrich № 40,833-6). ZnO nanoparticles with a size of 20 nm (Russian local supplier); SiO2 nanoparticles with a size of 14nm ( Aldrich No. 066K0110) were used for structuring nanomodification. 2,2-Dimethoxy-2-phenylacetophenone (Aldrich No.19,611-8) and Bis (5-2,4-cyclopentadien-1-yl) bis[2,6-difluoro-3-(1H-pyrrol-1-yl)phenyl]titanium (Irgacure 784) were used us initiators of the photopolymerization.

Transmission spectra were measured on a spectrophotometer Perkin-Elmer 555 UV-Vis. For the IR spectra we used Fourier IR spectrometer FSM 1201 Manufacturer Company "Monitoring". Samples were prepared by pressing pellets with KBr. The index of refraction is determined by the Maxwell-Garnett effective medium theory [1] and by Abbe refractometer.

We study sorption of water vapor by gravimetric method. Hardness is measured by Brinell method with "Bulat-T1" device. Light scattering is measured by the photometric sphere method in accordance with the recommendations of the European standard ASTM D1003.

Investigation of the surface profile of samples was made with an atomic-force microscope Ntegra in contact mode.

### 2.2. Preparation of monomer solutions and films of the nanocomposites

We find some effects at mixing of nanoparticles with monomer mixture. At nanoparticles introduction in the composition of the monomer BisA/2Carb (30/70) above a concentration of 8 wt. % the viscosity of solutions increases greatly. At higher concentrations (14 wt. % ZnO and 10 wt. % SiO2) the monomer mixture becomes very viscous transparent gel at room temperature.

Transparent nanocomposites were obtained up to 14 wt. % of ZnO and 12 wt.% SiO2 nanoparticles with formation of transparent film after UV-curing. More than 14 wt. % of ZnO and 12 wt.% SiO2 nanoparticles addition to the monomer mixture resulted in turbid films. The system becomes heterogeneous.

Polymer films (thickness from 12 to 100 microns) were obtained from the previously prepared solutions containing monomer, nanoparticles and initiator. The drop of solution is trapped between two polyester films to prevent inhibitory effects of oxygen. All experiments were accomplished at room temperature in air without special inert atmosphere. UV curing was made by a mercury lamp (100 W) used at the mercury line at 365 nm.

## 2.3. Results

The films are transparent in the visible and UV spectral region. A significant decrease in optical transmission is observed at high concentrations of ZnO (more than 14 wt. %) and concentrations of SiO2 more than 8 wt.%.

The values of the refractive index nanocompositions shown on Figure1.

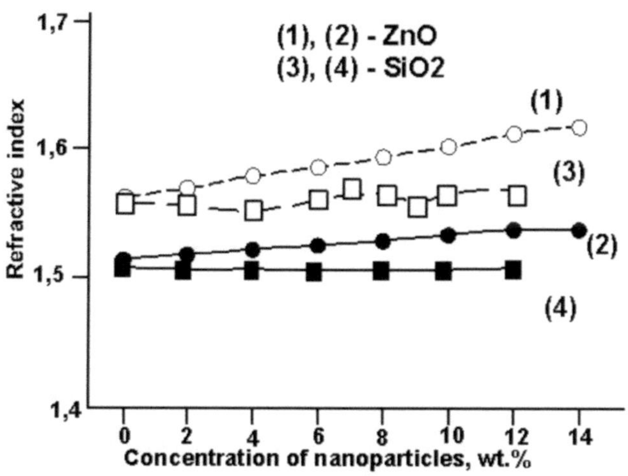


**Figure 1.** Refractive index is determined by the Maxwell-Garnett effective medium theory (3), (4) and by Abbe refractometer (1), (2)

For the refractive index were used effective medium model of Maxwell-Garnett:

$$\frac{\varepsilon_{eff} - \varepsilon_2}{\varepsilon_{eff} + 2\varepsilon_2} = f_1 \frac{\varepsilon_1 - \varepsilon_2}{\varepsilon_1 + 2\varepsilon_2} \qquad (1)$$

$\varepsilon_1$ – permittivity of the medium;
$\varepsilon_2$ – permittivity of inclusions;
$\varepsilon_{eff}$ – permittivity of the composite medium;

$$f_1 = \frac{1}{V} \Sigma_i V_i \qquad (2)$$

volumetric filling factor; ($V_i$ - volume of i-th particle, V - volume of the composite environment).

This model is applicable when the volume filling factor: $f_1 \leq \frac{1}{3}$, i.e. fraction of inclusions is small. The obtained values for the film samples by Abbe are greater then calculated by Maxwell-Garnett model. Model of Maxwell-Garnett is applicable in the case if nanoparticles are distributed inside any matrix without interaction between nanoparticles and polymer. According our recent results, there are interactions between nanoparticles and polymer matrix. Therefore, the use of this model is not entirely correct. In addition, the refractive index obtained by Abbe of the film samples are greater than calculated by Maxwell-Garnett model, as we assume due to the fact that the polymerization did not pass until the end and the residual monomer are contain in the matrix. With the introduction of the maximum possible concentration of SiO2 (12 wt.%), the refractive index of the composition is reduced by 0.02 compared with the initial monomer mixture. Refractive index of composition with maximum ZnO concentration (14 wt. %) is increased by 0.045 compared with the polymer without nanoparticles.

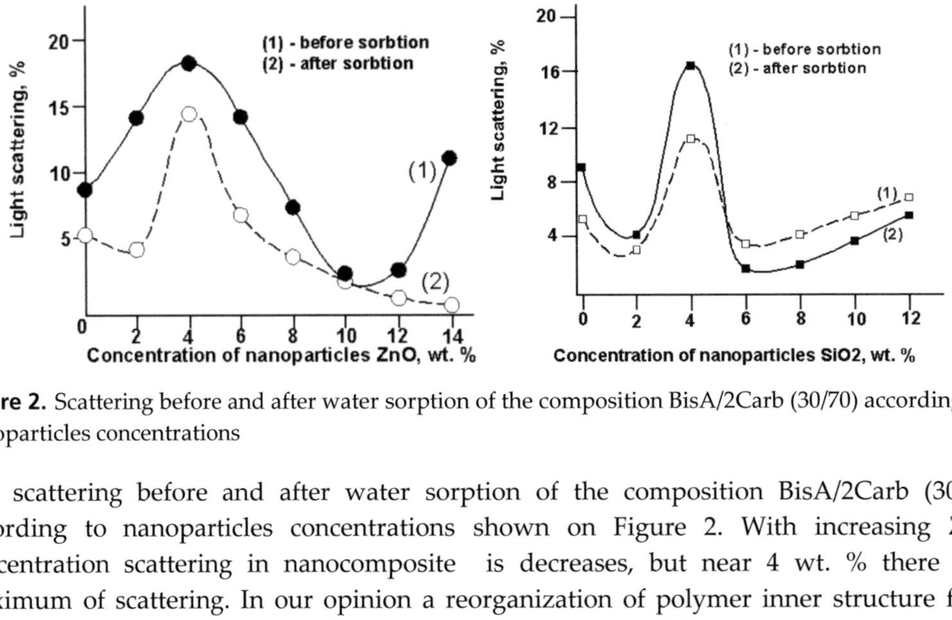


**Figure 2.** Scattering before and after water sorption of the composition BisA/2Carb (30/70) according to nanoparticles concentrations

The scattering before and after water sorption of the composition BisA/2Carb (30/70) according to nanoparticles concentrations shown on Figure 2. With increasing ZnO concentration scattering in nanocomposite is decreases, but near 4 wt. % there is a maximum of scattering. In our opinion a reorganization of polymer inner structure from polymeric structure with inclusion of nanoparticles to self-organized nanocomposite structure is occur near this concentration. When the concentration of ZnO is in the range from 8 to 14 wt%, scattering is almost independent on the concentration of nanoparticles. 8 - 14 wt% ZnO nanoparticles is sufficient for uniform distribution inside volume as a result homogeneous polymer nanocomposites are formed. In this nanocomposite structure, the scattering decreases compared to pure polymer, approximately twice. After the water sorption tests, the concentration dependence of the scattering of remains, but for all concentrations of the magnitude of sorption decreases.

Dependence of scattering of nanocomposites before and after the water sorption according to the SiO2 nanoparticles concentration is shown in Figure 2.

Due to lack of nanoparticles concentration in the field up to 8 wt % the polymer is not homogeneous, resulting in a nonmonotonic variation of the scattering is observed. At concentrations higher than SiO2 8 wt%, nanoparticles are uniformly distributed throughout the volume and the quasi-homogeneous composite is formed. The scattering in the new structure is reduced compared to the polymer matrix approximately reduced by about half. After the water sorption tests the scattering dependence is conserved. For all compositions, except for pure composite and the composite with the addition of 12 wt.% SiO2, the scattering is qualitatively does not change, but its value decrease. This phenomenon is explained by the possible water plasticizing effects, for example, see [2,3].

The water sorption experiments were conducted to study changing the internal structure as a result of the introduction of nanoparticles. As can be seen from the figure 3, the sorptions of nanocomposites have strong dependence from nanoparticles concentrations. There is a sorption maximum near 4 wt % ZnO (reorganization of polymer inner structure). At ZnO nanoparticles concentration above 12 wt. % water sorption increase, perhaps as a result of disordering of the nanocomposite. For pure polymer value, water sorption is 23 %. The introduction of 10 wt. % ZnO nanoparticles achieved reduction in water sorption by 5 times compared with the pure composition.

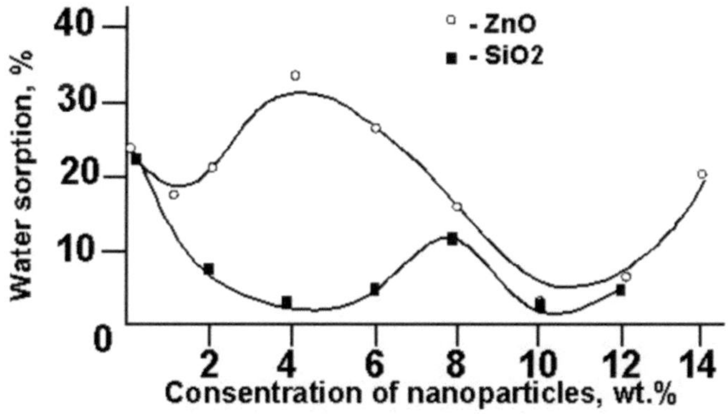


**Figure 3.** Water sorption BisA/2Carb(30-70) composites according to nanoparticles concentrations

With increasing SiO2-concentrations higher than 12 wt.% the vapor sorption decreases steadily, reaching 2.5%. As compared to the unmodified polymer, the introduction of SiO2 nanoparticles reduces the water absorption ten times. Formation of the extremum of adsorption at a concentration of 8 wt.% SiO2 found no explanation.

The investigations of film hardness is an indirect way to study the effect of nanoparticles on the structural change of nanocomposites (Figure 4 ).

With the introduction of ZnO nanoparticles the film hardness decreases and remains almost unchanged until to 10%. Further, the hardness of the film increases and reaches the value of pure polymer. With the introduction of 2 wt.% SiO2 film hardness increases sharply in comparison with the original. In the concentration range from 2 to 8 wt. % the composite film hardness is decreases sharply. Further, with increasing SiO2 concentrations hardness values start to increase and at the maximum concentration of SiO2 (12 wt. %) it become comparable with the value of the pure polymer.

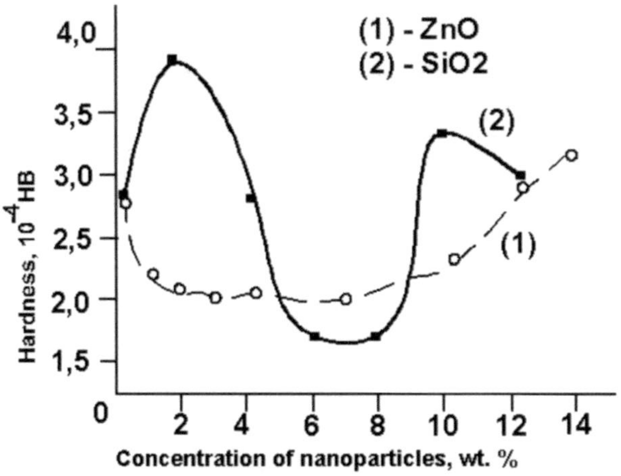


**Figure 4.** Hardness of BisA/2Carb(30-70) composites according to nanoparticles concentration

The above changes in the properties of polymer nanocomposites (water sorption, light scattering, hardness) can be explained by a modification of the supramolecular structure of the polymer as a result of possible interactions of nanoparticles with active groups of the monomers [4].

Nanoparticles influence on the polymer structure was confirmed by the investigation of the surface relief and rigidity of the surface nanocomposites films made atomic force microscopy (Figures 5 and 6).

**Figure 5.** AFM. 5×5 mkm. Relief of surface ZnO polymer films

As can be seen from Fig.5, there are essential changes in the composition structure as compared with the original by introduction 4 wt. % ZnO nanoparticles. The formation of separate ZnO structured polymer regions were observed clearly. The graininess structure are observed throughout the volume of material when ZnO concentration are achieves 10 wt. %.

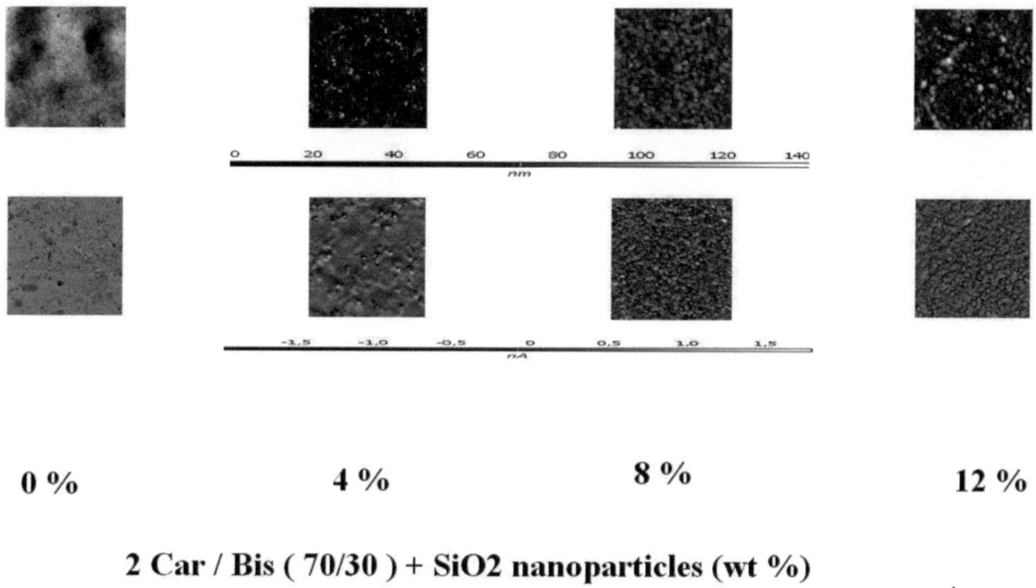


**Figure 6.** AFM. 5×5 mkm. Relief (top) and rigidity of surface SiO2 polymer films (bottom)

As can be seen (Figure 6), already with the introduction of 4 wt.% of SiO2 nanoparticles, significant changes in the nanocomposition structure are observed as compared to the pure polymer structure. The formations of separate domains are observed clearly. With the introduction of 12 wt.% SiO2, structure of uniform graininess is formed throughout the material.

Apparently, formation of nanocomposition structure is due to the ability of nanoparticles to create on the surface bonds with the active groups of the monomer molecules and to act as centers of polymerization. When the ZnO concentration is more than 10 wt. % (SiO2 concentrations of more than 8 wt.%) free polymer phase disappears and all the monomer is consumed for the formation of polymeric spheres on the surface of nanoparticles.

Thus, at low concentrations of nanoparticles, modified polymer areas are still small and composites are heterogeneous. Heterogeneous structures are reflected in their properties. Increasing nanoparticles concentration is lead to increasing hybrid field's amount and its size, so uniform nanocomposite structure begins to form. The result is the formation of submicron spheres around each nanoparticle and quasi-homogeneous material is formed. Indeed (Figures 5 and 6), submicron spheres formed around each nanoparticle possess almost identical diameters. This fact can be explained by identical growth rates of these spheres. As a result, a structure consisting of spherical particles forms a self-organized quasi-lattice. Eventually, above-named effects leads to a homogeneous distribution of nanoparticles and homogeneous composite environment are formed.

Possibility and mechanism of polymerization on surface of nanoparticles were investigated by FTIR (Figures 7, 8).

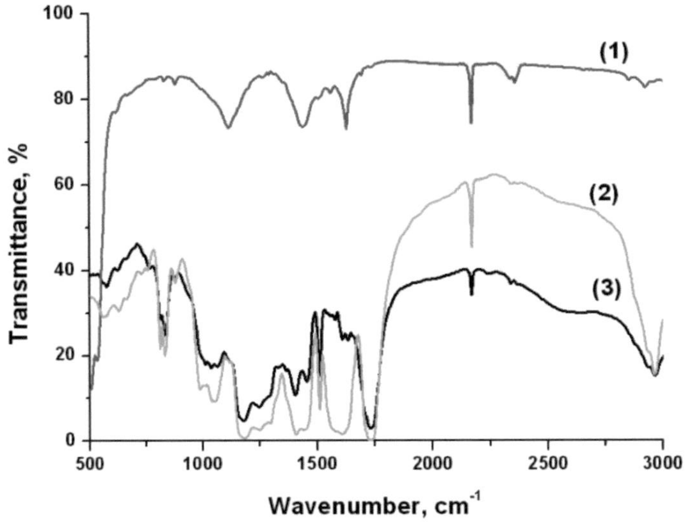


**Figure 7.** (1) - Nanoparticles ZnO; (2) - composition 2Carb/BisA (70/30) + 12 wt. % ZnO; (3) - 2Carb/BisA (70/30)

In the FTIR of BisA/2Carb (30/70) compositions the bands with maximum absorption at 1737 cm-1, 1410 and 1188, 1050 cm-1 were observed. The first band is characteristic for the stretching vibrations of the carboxyl group, the next three due to a combination of plane deformation vibrations of the hydroxy group and the stretching vibrations of the C-O in carbonic acids [5].

Intense peak around 500 cm-1 belongs to the Zn-O vibrations in the ZnO crystal [6]. In the IR spectra of the nanocomposite the band at 1720 cm-1 (group C = O) is observed and the 1410-1450 cm-1 bands corresponding to symmetric stretching vibrations of the carboxylate anion are greatly enhanced.

It is important to note the following. The band 500 cm-1, relating to the Zn-O vibrations in the crystal ZnO, is preserved. The absorption band 1620-1550 $cm^{-1}$ appear in the composite. This region is characteristic of asymmetric stretching vibrations of the carboxylate anion. As reported in [7 - 9], the interaction of inorganic nanoparticles (ZnO) and carbonyl groups can cause changes in the IR, because the metal atoms can be taken electron pair of the carbonyl oxygen.

The increase in the absorption band of 1600 cm-1 may be due to the complexation between the polymer and nanoparticle.

In addition, the band 1640-1650 cm-1 is characteristic of the C = C stretching vibrations in the CH2 = CHR, and the band 990 cm-1 is characteristic for C-H bending vibrations in the CH2 = CHR [10]. The appearance in the spectra of the composites above the bands confirmed that the group CH2 = CHR formed. These bands may be formed during the polymerization of the composite only on the surface of nanoparticles of zinc oxide when the latter acts as a photocatalyst.

Principal scheme of described photopolymerize process is shown in Figure 8.

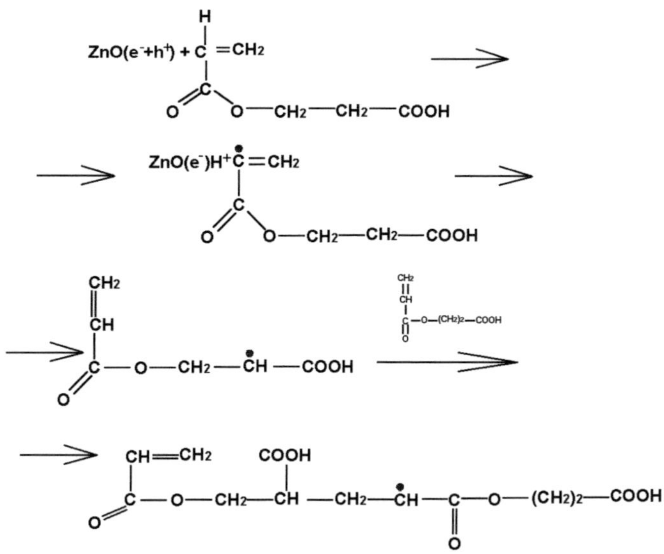

**Figure 8.** Scheme of photopolymerize process

The SiO2 nanoparticles can form bonds with the monomers and act as centers of polymerization the same way. It was confirmed by FTIR spectroscopy (Figure 9). The peaks 500 cm-1 and 1109 cm-1 were attributed as the stretching vibrations of Si-O-Si bounds on SiO2 surfaces [10]. The 1061 cm-1 band is attributed as the C-O stretching vibrations in the C-O-H [10]. The degradation of strong bands 471 cm-1 and 1107 cm-1 and the emergence of new band 1061 cm-1 in the nanocomposites may be indicative of polymerization on the surface of nanoparticles due to the interaction of COOH-groups of the monomer with nanoparticles SiO2.

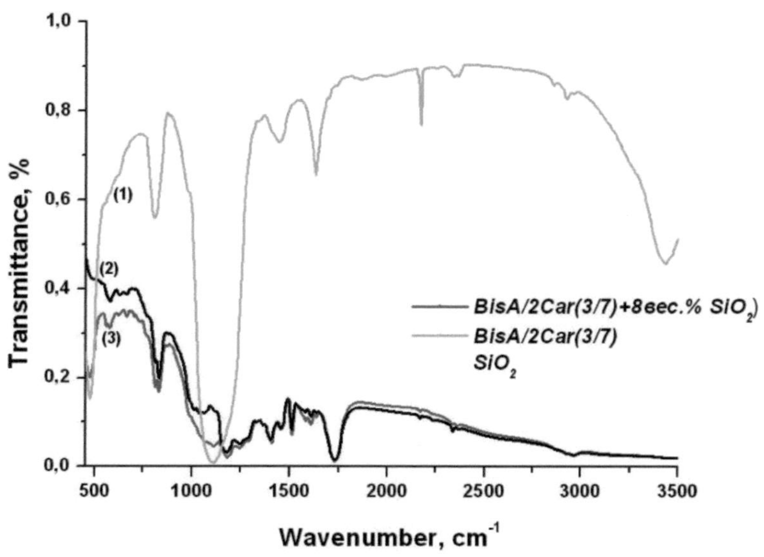


**Figure 9.** FTIR spectra: (1) - SiO2; (2) - BisA/2Car (30/70) + 8 вес. % $SiO_2$; (3) - pure polymer BisA/2Car (30/70)

## 2.4. Discussion results

According to the previously mentioned nanoparticles can be involved in UV polymerization process by formation of bonds between nanoparticle surfaces and carbonyl polymer group. Thus, the nanoparticles act as formation centers of a new polymer phase - the nanocomposite possessing different properties in comparison with the unfilled polymer. According to FTIR spectra of nanocomposites we assume that nanoparticles are participated in polymerization processes and act as a photocatalysts. Our hypotheses is supported by the formation of micron size spheres in nanocomposite around each nanoparticle (Figure 5, 6).

Nanosized semiconductor clusters have the potential to photooxidation. Photocatalysis take place through the combined effects of photoelectrons production at UV light absorption and high surface area in that electron transfer induced polymerization.

Nanosized semiconductor clusters may participate in catalysis of redox - processes on surfaces exposed to light, that is, they can be photocatalyst [11,12]. This is due to a combination of effects: formation of photoelectrons by absorption of UV light and, in addition, a large specific surface of nanoparticles contributes to their high catalytic activity. Thus, when the semiconductor catalyst absorbs the photon with energy equal to or greater than the value of the band gap, an electron from the valence band can move into the conduction band to form electron-hole pairs. In the future, such a pair can take part in the reactions of donor-acceptor mechanism on the catalyst surface, that is, can begin the process of polymerization. According to the work [13], the UV illumination of a semiconductor photocatalyst activates the catalysis and accelerates establishing a redox environment in the aqueous solution. Semiconductors act as sensitizers for light induced redox processes due to their electronic structure, which is characterized by a filled valence band and an empty conduction band [14]. So, the photocatalytic process of photopolymerization was induced as a result of light absorption on the surface semiconductor nanoparticles. The work [12] describes the process of polymerization of methylmethacrylate initiated by TiO2 nanoparticles.

In our experiments we observed a similar process that proves by AFM and FTIR (the formation of equal spheres around the each nanoparticle, which are the center of polymerization and creation of chemical bonding between nanoparticle surface and polymer). Formation of spheres around each nanoparticle during photopolymerization on its surface appears to change a transformation of polymer structure. All nanocomposite properties exhibit extrema in the same areas of concentration of nanoparticles. Characteristic ranges of concentrations are 4 and 10 wt % for ZnO and 4 and 8 wt % for SiO2. According AFM photos in this concentration range, new phase - nanocomposite are generated on nanoparticles surface with increasing amount nanocomposite in volume. This patterning affects the changes in material properties, light scattering and a water sorption are decreasing significantly.

At nanoparticles concentration more than 8 wt % for SiO2 and 10 wt % for ZnO all monomer mixture will involved in formation of nanocomposite phase, perhaps next increasing of nanoparticles concentration will result on competition between nanoparticles as the center of polymerization. At 8 wt % for SiO2 and 10 wt % ZnO AFM photo show structures consisting from micro - spheres occupying all volume. In this concentration range light scattering as well as water sorption increasing. The above considerations explain the extrema of properties with increasing concentration of nanoparticles.

## 3. Holographic writing of periodic lattice in nanocomposite material

Holographic writing was conducted when recording the periodic structures in an interference field created by the interaction of two plane waves with wavelengths of 325 nm and 442 nm. The structures period was 2 microns. The diffraction efficiency was determined at a wavelength of 633 nm as a ratio of the first-order diffraction intensity to the incident radiation intensity. The layers were formed on a glass substrate in the gap between the glass and a polyester film. The thickness of the layer depended on the size of the filling and was 20 microns.

Abbreviation of used chemicals: 2-carboxyethyl acrylate (2Carb); Bisphenol A glycerolate (BisA); 2-Phenoxyethyl acrylate (PEA).

Tree methods of post exposition processing were used:

1. After exposure by interference field any additional processing was not made.
2. After exposure by interference field nanocomposite film were developed by isopropyl alcohol ablution.
3. After exposure by interference field were made uniform UV-radiation exposure.

Holographic writing on nanocomposites is due on light induced nanoparticles redistribution. Effects of light induced nanoparticles redistribution in nanocomposite is a new effect discovered recently. It takes place at photopolymeric nanocomposite irradiation by periodic light distribution, for example by lattice made by interference of two laser beams.

The first time these processes were found by Tomita and co-workers in 2005 on organic–inorganic nanocomposite photopolymer system in which inorganic nanoparticles with a larger refractive index differs from photopolymerized monomers are dispersed in uncured monomers [15]. Inorganic materials possess a wide variety of refractive indices that give us the opportunity to obtain much higher refractive index changes $\Delta n$ than conventional photopolymers, while maintaining low scattering losses [16].

Explanation of effects was made in the work Y. Tomita at all [17]. For monomers with radical photopolymerization spatially nonuniform light illumination will produce free radicals by dissociation of initiators, and subsequent reaction of free radicals with monomers, which leads to chain polymerization of individual monomers in the bright regions.

This polymerization process lowers the chemical potential of monomers in the bright regions, leading to diffusion (short distance transportation) of monomers from the dark to the bright regions. On the other hand, photosensitive inorganic nanoparticles have diffusion from the bright to the dark regions, as illustrated in Figure 10 left, since the particles are not consumed and their chemical potential increases in the bright regions as a result of the monomer consumption. Such a mutual diffusion process essentially continues until the monomers are consumed completely by the monomolecular and bimolecular termination processes and until the high viscosity of a surrounding medium consisting of polymerized monomers makes monomers and nanoparticles immobile. As a result the spatial distribution of nanoparticles is also fixed and a refractive-index grating is created as a result of compositional and density differences between the bright and the dark regions [17].

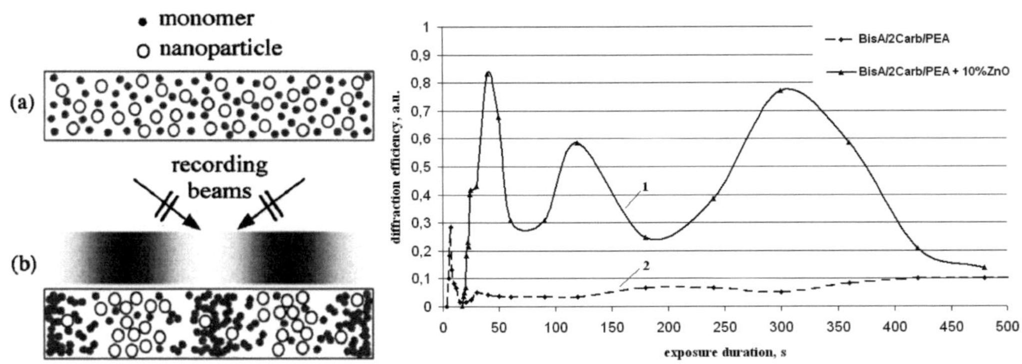


**Figure 10.** In the left - nanoparticles transportation in photopolymer according to work [18], in the right - our experimental diffraction efficiency dependence on exposition with and without nanoparticles.

Figure 10 shows diffraction efficiency dependence of photopolymerizable nanocomposite material from exposition time and nanoparticles concentration. Unlike of classic holographic photopolymer in our material unpolymerized materials were dissolved by alcohol and removed after exposition (method 2). According to schema (Figure 10 left) unpolymerized monomer in dark areas was removed by dissolution. High augmentation of diffraction efficiency at 12% nanoparticles concentration is a result of nanoparticles redistribution.

Detailed investigation of temporal traces of the first order diffraction efficiencies DE for the nanocomposites samples give oscillation of DE at increasing of exposition time and its dependence from processing of the samples. Figure 10 left shows the diffraction efficiency dependence from exposition for the nanocomposite BisA/2Carb/PEA 25/55/20, ZnO 10% in comparison with the monomer composition based of the same components without nanoparticles (Energy density: $2 \cdot 10^{-2}$ J/cm$^2$, a period of structures: 2 microns). We see a significant increase in the diffraction efficiency for the nanocomposite.

Figure 11 shows diffraction efficiency dependence from exposure and its change during one day storage and after uniform UV-radiation exposure (365 nm, $3 \cdot 10^{-1}$ J/cm$^2$) (method 3).

The periodic nature of the kinetic curves can be seen. The first maximum of diffraction efficiency just after exposure (Figure 11, curve 1) at low exposures can be connected to increasing of refractive index of composition after polymerization. Refractive index are increasing on 0,04 in result of photopolymerization In bright areas. In dark areas photopolymer will keep in liquid state. Difference of RI between liquid and solid (polymerized) composition will result in increasing of DE. After entire polymerization either by long exposition or by UV light 365 nm, whole composition will be polymerize and DE will decrease (curve 2, Figure 11).

At the same time nanoparticle diffusion process take place. It is seen that DE builds up after a relatively long induction time period that usually corresponds to the time duration to consume contaminated oxygen [18]. In fact nanoparticles displacement will begin after at last 50% photopolymer conversion that result on induction period. If uniform exposition of nanocomposite after laser lattice writing take place, whole composition will be polymerized and refractive index modulation of cured material will be connected to displacement of components only.

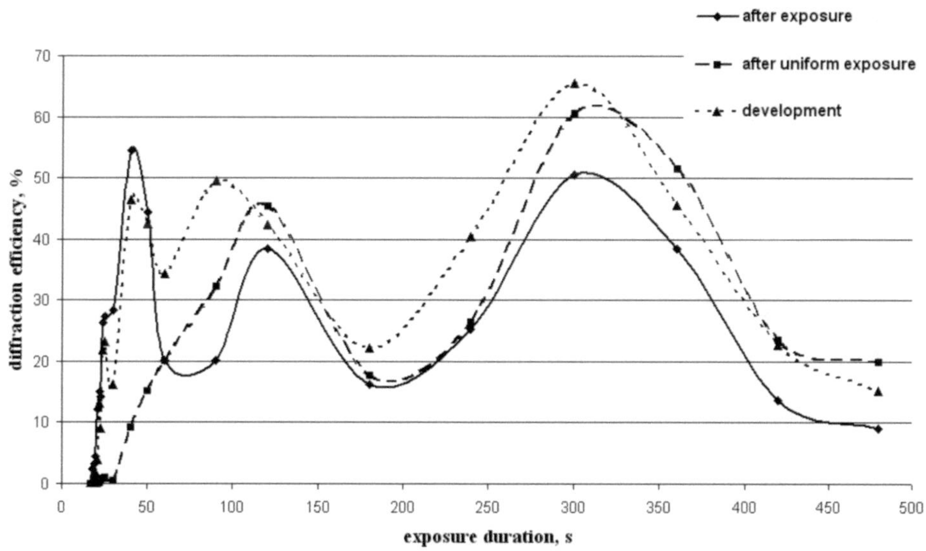


**Figure 11.** The dependence of the diffraction efficiency from exposure duration for the nanocomposite BisA/2Carb/PEA 25/55/20, ZnO 10%. Just after exposure (1), after uniform exposure (2) and one day after exposure (3). Energy density: $10^{-2}$ J/cm$^2$, a period of structures: 2 microns

According to the work [19] DE builds up after a induction time period, going to the maximum and finally decays almost to zero, which is a typical behavior of one component photopolymers without any binder materials. Our nanocomposite have no conventional binder, so its exposition curves should have the maximum, but unlike of conventional photopolymer, nanocomposite having two diffusing comments, i.e. monomer and nanoparticles form two maximums. Mitual location of the maximums depend on diffusion coefficients of monomer and nanoparticle.

Stability of obtained DE at last during one day proves components diffusion nature of the maximums at 120 and 300 s (Figure 11). After uniform exposition by 365 nm UV, nanocomposite become solid, diffusion processes will stop and will keep existing distribution of nanoparticles in composition. Some increasing of DE after exposure can be due to compression of nanoparticles in dark areas in solid nanocomposite.

Application of more viscous composition will result of diffusion braking and can be expect to braking of back diffusion at overexposure resulting of DE fall after maximum (Figure 11).

For that reason we use $SiO_2$ based nanocomposite very viscous and receive absence of DE fall after maximum (Figure 12).

Instead of lattice writing in ZnO based nanocomposite (Figure 12), DE increase here up to maximum value and keep stable after exposition. Lattice made in nanocomposite is stable and survive heating up to 150C during one hour without fall of DE.

DE of material is result of nanoparticles displacement, so nanoparticles moved to the dark areas can be made visible by different microscopy methods.

Transmission optical microscopy photo on micro- cut and confocal microscopy photo of micro lattice are shown in Figure 13. According Figure 13 (right) there takes place short distance nanoparticles transportation to the bright regions with change of solubility of material and reinforcement of polymerized 3D lattice formation.

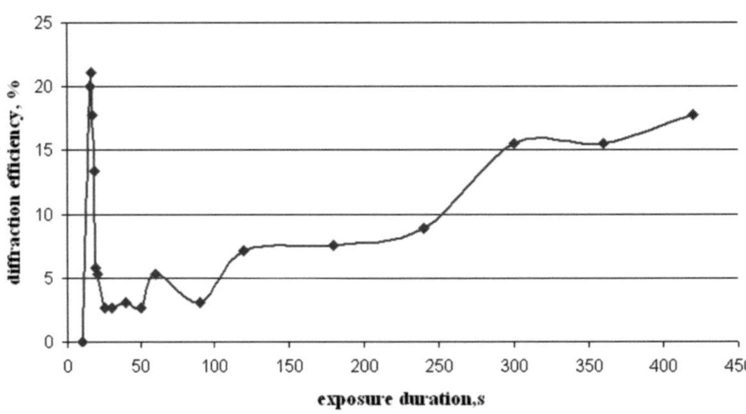


**Figure 12.** The dependence of the diffraction efficiency from exposure duration for the nanocomposite BisA/2Carb 30/70, $SiO_2$ 6%. Energy density: $10^{-2}$ J/cm$^2$, a period of structures: 2 microns

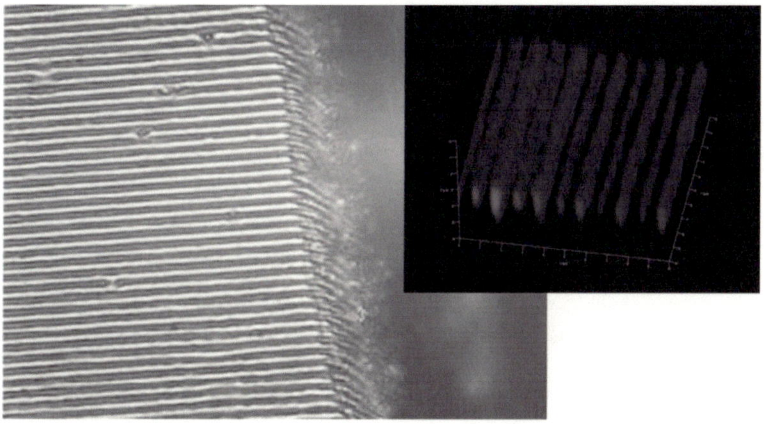

**Figure 13.** Micro- photo of cut of 3D structure (objective 160x, aperture 1,4) and confocal microscope plot of same structure (insert)

Next show the first experiments on making visible nanoparticles redistribution by removing of polymer without nanoparticles by ion etching. Result of ion etching of previously made holographic micropatterns in nanocomposite is shown on Figure 14.

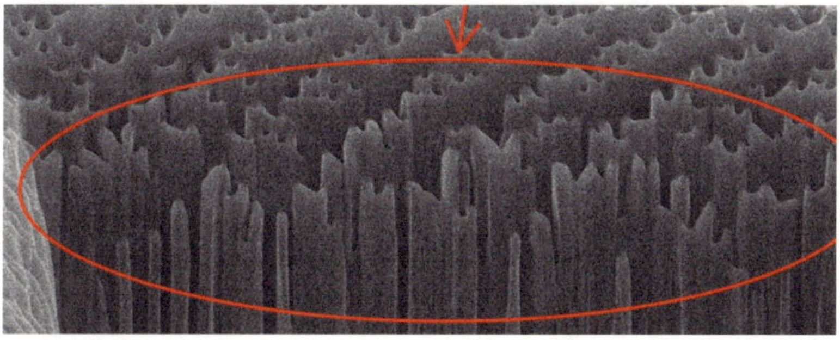

**Figure 14.** Ion beam etching of micropatterns will result on evaporation of polymer and keeping of nanocomposite areas with formation of vertical elements 100 nm width and 10 nm height

We think that ion beam etch free polymer and to not touch nanoparticles enriched areas. In result after that nanoparticles redistribution becomes visible, it is forms subwavelength nano- sized columns with high aspect ratio.

Direct observations of nanoparticles displacement were made by dissolution of liquid uncured composition and surface investigation by AFM method.

Figure 15, c show AFM surface of nanocomposite cured by uniform UV exposition. It is seem as nanoparticles distributed on whole surface.

After lattice writing and dissolution of uncured composite we obtain AFM photo (Figure 15 a, b). Clear visible concentration of nanoparticles in dark areas. Nanoparticles arrange maximal densely and have no in light areas.

So we find direct confirmation of light assistant nanoparticles displacement in nanocomposite.

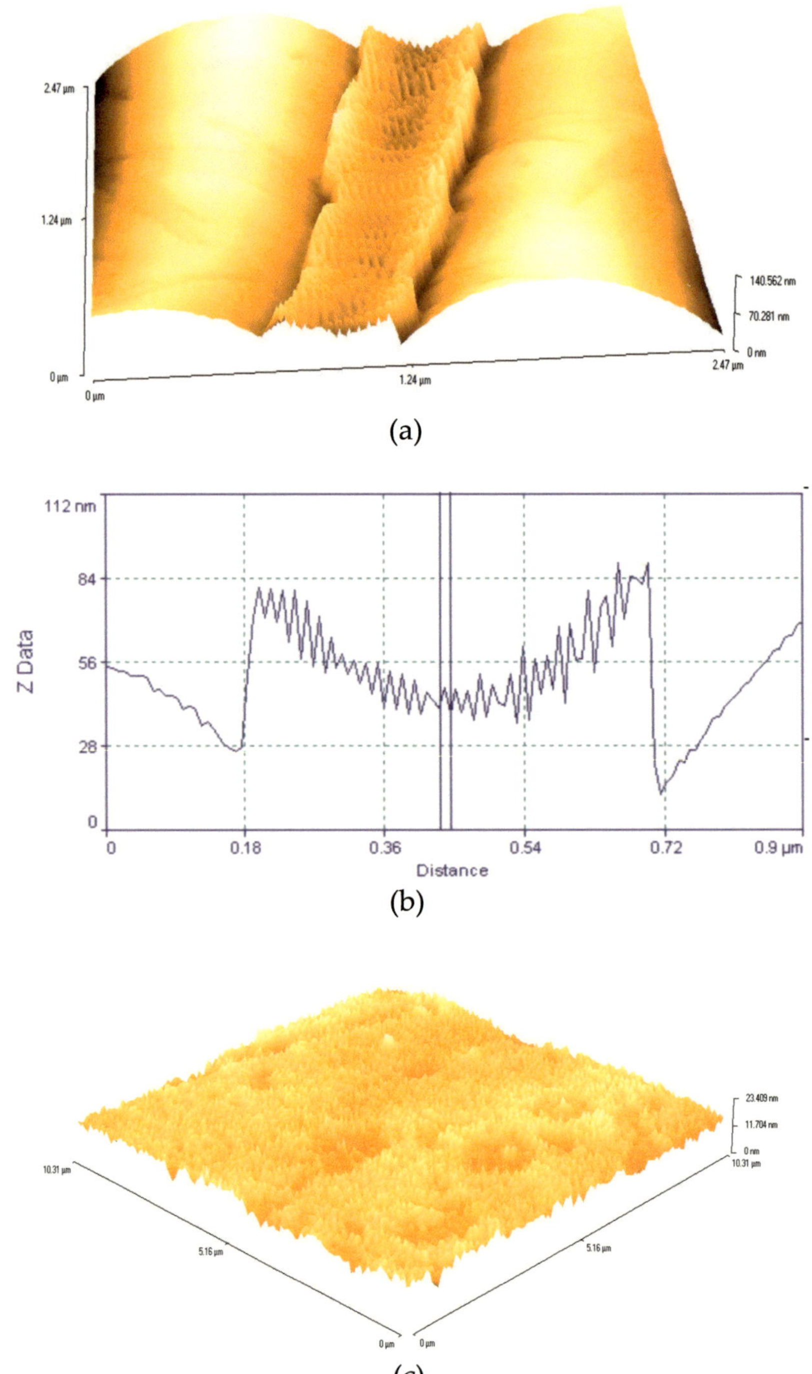


**Figure 15.** 3D image (a) and plot (b) of a fragment of a polymer structure recorded on the nanocomposite BisA/2Carb/PEA 25/55/20, ZnO 10% and the structure of films of the nanocomposite (c). Period of structures: 2 microns

## 4. Self-organization processes in photopolymerizable nanocomposite

The main problem of current photolithography is diminishing of minimal feature sizes up to subwavelength value. The smallest feature size $X_{min}$ that can be projected by a coherent imaging system is $X_{min}=\lambda/2NA$, and the depth of focus DOF is $DOF=\lambda/[2NA^2]$, where $\lambda$ is the wavelength of the illumination and NA is the numerical aperture. The most ordinary way to attain smaller feature sizes is to reduce the wavelength up to excimer laser wavelengths (248 or 193 nm). The NA is typically between 0,5 and 0,8, so the feature size is on the order of the exposure wavelength.

A unique way to overcome diffraction limit of resolution is to use non-linear light transformation in special photoresist. For the same reason special UV-curable nanocomposite with strong non-linear and self-writing effects overcoated on photoresist to improve light distribution in the spot can be used.

Same proposed technique is applicable for deep lithography based on 365 nm UV light with high scattering to improve shape of small feature in results of geometrical optical laws perturbation at use light self-focusing in materials. If to make nanocomposite system with self-writing effects and place it as a topcoat, we will obtain self-writing subwavelength artificial waveguide that will guide the light to small subwavelength spot on photoresist surface. So to make it it's necessary to develop special material with self-writing effects.

Light self-focusing and self-organization effects at UV curing of acrylate based nanocomposites were investigated previously [20].

There are a few effects:

1. *Self-focusing of the light in material with positive refractive index change at photopolymerization*

Effect of light self-focusing in optical material having proprieties of positive change of refractive index (RI) at light action is widely investigated recently. For example, in cited work the results of light self-focusing and self-written waveguide preparation process obtained on glass light sensitive materials are summarized [21].

Our experiments show important influence of well known oxygen inhibition action on reinforcement of self-focusing light in photopolymer. Oxygen inhibition action to acrylate photopolymerization described previously [18].

2. *Short distance nanoparticles transportation*

Effect of light induced nanoparticles redistribution in nanocomposite is a new effect discovered recently. It takes place at photopolymeric nanocomposite irradiation by periodic light distribution, for example by lattice made by interference of two laser beams.

The first time these processes were found by Tomita and co-workers in 2005 on organic–inorganic nanocomposite photopolymer system in which inorganic nanoparticles with a larger refractive index differs from photopolymerized monomers are dispersed in uncured monomers [15].

Explanation of effects was made in the work [17]. We use these effects to improve distribution of photopolymer at its curing in light spot. Next show results obtained at microstructure writing by projection lithography.

Self-writing and self-focusing effects discussed above are applicable for diminishing polymerized area initially corresponding to light distribution in objective spot as well as to overcome geometric distribution of the light in focus. The main effect is light self-focusing that can be reinforced by oxygen inhibition [22, 23] and nanoparticles redistribution at photopolymerization. Figure 16 shows proposed application of self-writing processes in projection photolithography: using nanocomposite with self-writing effects overcoated on photoresist to improve light distribution in the spot on the photoresist surface.

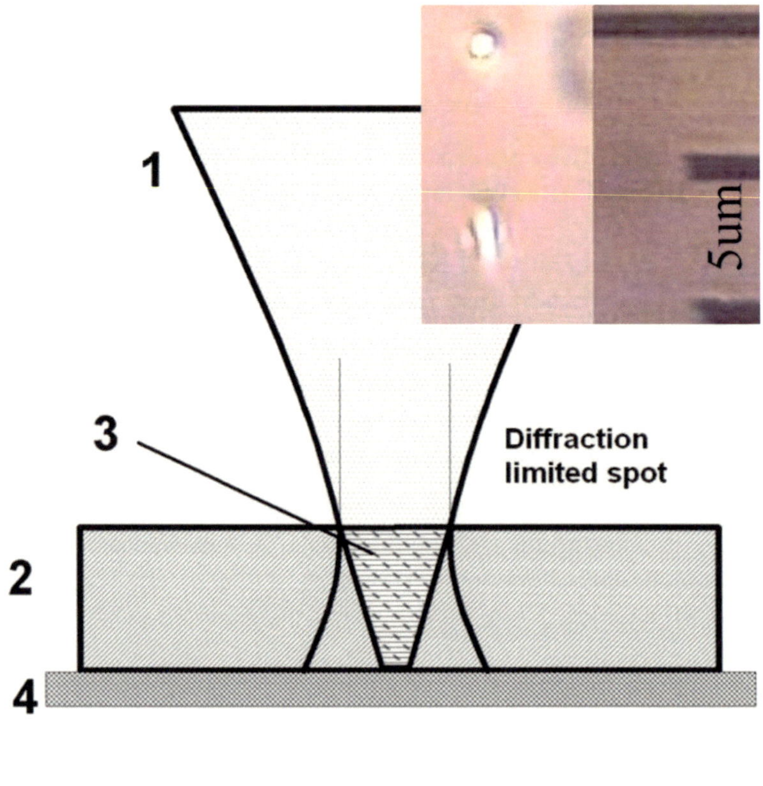


**Figure 16.** Model of light redistribution behind of lens spot in result of self-writing processes.1 - UV light beam, 2- nanocomposite, 3- Tip self-writing, 4- photoresist. Polymerized small one micron sized cylinder with vertical borders made by this process (insert).

Two monomer compositions (with and without nanoparticles) were used. In Figure 17 are represented polymer microstructures obtained in case of 2-carboxyethyl acrylate and bisphenol A glycerolate composition (without nanoparticles). Relation between dimensions/structure shape and exposition is observed. In this experiment was used projection of spot expected diameter 1, 2, 4, 6 um. Composition without nanoparticles doesn't allow to obtain structures with all expected diameters. All cylinders based on this composition were at least 3-4 um diameter; in case of 1 um expected spot size formation of elements didn't take place.

Elements form tends to the cone that corresponds to energy distribution in spot.

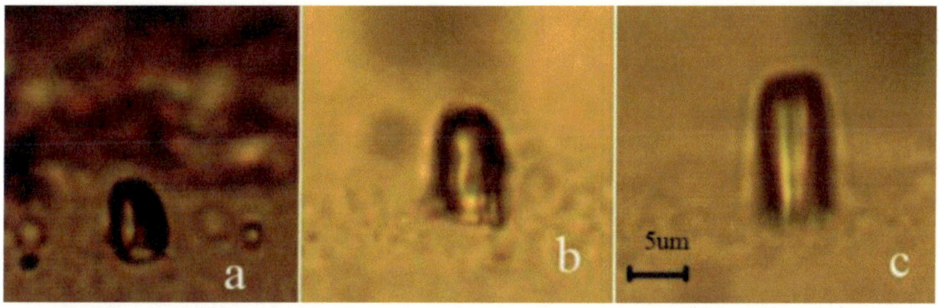


**Figure 17.** Structures obtained from photopolymer without nanoparticles: a – diameter 3 μm (spot size 2 um); b –diameter 4 um (spot size 4 um); c –diameter 6.5 um (spot size 6 um)

In contrast with this series, experiments with nanocomposite gave different results – we made structures with expected diameters 1 um and obtained diameter less than 1 um with form near to cylinder Figure 18.

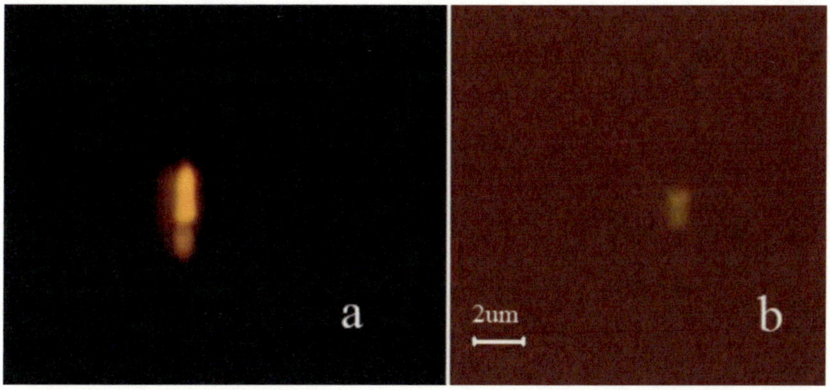


**Figure 18.** 2 um-height structures obtained from nanocomposite: a – diameter 0.7 um (spot size 1,5 um), b – diameter 0.6 um (spot size 1,5 um)

Figure 18 shows that use of nanocomposite give formation of subwavelenght elements formed with geometric optics law violation (formation of self-writing cylinder smaller in comparison to spot size two times). In fact used lens should form cone 20 degrees, but in result of nanocomposite self-organization it is form cylinders with vertical borders and size two time less than calculated one, that confirm our guess-work on self-organization effects in nanocomposite.

## 5. Conclusion

Nanocomposite UV-curable material having set of self-writing proprieties such as: light self-focusing and light induced nanoparticles redistribution were development. The sorption of water vapor, Brinell hardness, optical transmission, refractive index and light scattering of film polymer ZnO -nanocomposites were studied. Composites are transparent in the visible area at high concentrations of ZnO nanoparticles (14 wt. %). With the introduction of 14 wt % ZnO refractive index increases by 0.045. With the introduction of 10 wt. % ZnO the sorption decreases by five times. Hardness, until a maximum concentration of nanoparticles 12 wt. % ZnO does not exceed the hardness of the pure polymer, while light scattering is not increased. Nonmonotonic changes in the properties, the AFM data and the IR spectra were explained the ability of nanoparticles to act as centers of polymerization and to form a granular structure in the nanocomposite

Nanocomposite material is applicable for subwavelength optical projection lithography and holography. Experiment on optical projection writing made at use of 0,2 aperture lens shows diminution of polymerized element smaller in comparison to spot size two times and transformation of initial conical light distribution to cylindrical one in result of self-organization processes in nanocomposite material. Holographic proprieties of material show high diffraction efficiency. Light induced nanoparticles distributions were investigated by different methods. AFM show redistribution of nanoparticles in dark areas.

## Author details

Igor Yu. Denisyuk*, Julia A. Burunkova, Vera G. Bulgakova and Mari Iv. Fokina
*Quantum Sized Systems, Saint-Petersburg National Research University of Information Technologies, Mechanics and Optics, Saint-Petersburg, Russia*

Sandor Kokenyesi
*Faculty of Science and Technology University of Debrecen, Debrecen, Hungary*

## 6. References

[1] Vinogradov A.P. (2001) Electrodynamics of composite materials. Moscow: Aditorial URSS. 176 p.
[2] Reitlenger S.A.( 1974) Polymer materials penetration. Moscow: Chemistry. 269 p.
[3] Prokofieva T.A., Davidova E.B., Kariakina M.I., Maiorova N.V (1980) Effect of adsorption of low molecular weight substances on the structure of crosslinked polyesters. VMS. (A) XXII. 1: 23-27.
[4] Jiguet S., Bertsch A., Judelewicz M., Hofmann H., Renaud P. (2006) SU-8 nanocomposite photoresist with low stress properties for microfabrication applications. Microelectronic Engineering. 83. 10: 1966–1970 p.

* Corresponding Author

[5] Silverstein, R.M.; Bassler G.C., Morrill, T.C. (1981) Spectrometric Identification of Organic Compounds. 4th ed. New York: John Wiley and Sons. 222 p.
[6] Liufu S.C., Xiao H.N.,. Li Y.P. (2005) Thermal analysis and degradation mechanism of polyacrylate/ZnO nanocomposites. Polym. Degrad. Stab. 87: 103-110 p.
[7] Lu X., Zhao Y., Wang C. (2005) Two-Polymer Microtransfer Molding for Highly Layered Microstructures. Adv. Mater. 17: 2481-2485 p.
[8] Lu X., Zhao Y., Wang C., Wei Y. (2005) Fabrication of CdS nanorods in PVP fiber matrices by electrospinning. Macromol. Rapid Commun. 26(16): 1325-1329 p.
[9] Bai J., Li Y., Zhang C., Liang X., Yang Q. (2008) Preparing AgBr nanoparticles in poly(vinyl pyrrolidone) (PVP) nanofibers. Colloids Surf. A. 329: 165-168.
[10] Kupcov A.H., Jijin G.N. (2001) Raman scattering and Fourier -IR polymers spectra. Moscow: Fismatlit.582 p.
[11] Hagfeldt A., Graetzel M. (1995) Light-induced redox reactions in nanocrystalline systems. Chemical Reviews. 1: 49-68 p.
[12] Dong C., Ni X. (2004) The Photopolymerization and Characterization of Methyl Methacrylate Initiated by Nanosized Titanium Dioxide. J. of macromolecular science Part A - Pure and Applied Chemistry. A41. 5: 547–563 p.
[13] Beydoun D., Amal R., Low G., McEvoy S. (1999) Role of nanoparticles in photocatalysis. J. of Nanoparticle Research. 1: 439–458 p.
[14] Hoffmann J. S., Pillaire M. J., Maga G., Podust V., Hübscher U., Villani G. (1995) DNA polymerase beta bypasses in vitro a single d(GpG)-cisplatin adduct placed on codon 13 of the HRAS gene. PNAS. 92 (12): 5356-5360 p.
[15] Suzuki N., Tomita Y., Kojima T. (2002) Holographic recording in TiO2 nanoparticle-dispersed methacrylate photopolymer films. Appl. Phys. Lett. 81. (2): 4121-4123 p.
[16] Suzuki N., Tomita Y. (2006) Highly transparent ZrO2 nanoparticle-dispersed acrylate photopolymers for volume holographic recording. Optics express.14. (26): 12712 – 12719 p.
[17] Tomita Y. and Suzuki N. (2005) Holographic manipulation of nanoparticle distribution morphology in nanoparticle-dispersed photopolymers. Optics Letters. 30. (8): 839-841 p.
[18] Andzejewska E. (2001) Photopolymerization kinetic of multifunctional monomers. Prog. Polymer Sci. 26: 605-665 p.
[19] Karpov G.M., Obukhovsky V.V., Smirnova T.N., Lemeshko V.V. (2000) Spatial transfer of matter as a method of holographic recording in photoformers. Optics Communications. 174: 391–404 p.
[20] Denisyuk I.Yu., Vorzobova N.D., Burunkova J.E, Arefieva N.N, Fokina M.I, (2011) Self-organization effects in photopolimerizable nanocomposite. Mol. Cryst. Liq. Cryst. 536. (1): 233-241 p.
[21] Monro T.M., Moss D., Bazylenko M., Martijn de Sterke C., Poladian L. (1998) Observation of Self-Trapping of Light in a Self-Written Channel in a Photosensitive Glass. Phys. Rev. Lett. 80. (18): 4072 – 4075 p.

[22] Denisyuk I.Yu., Fokina M.I., Vorzobova N.D., Burunkova Yu.E., Bulgakova V.G (2008) Microelements with high aspect ratio prepared by self-focusing of the light at UV-curing. Mol. Cryst. Liq. Cryst. 497: 228–235 p.

[23] Fokina M.I., Sobeshuk N.O., Denisyik I.U. (2010) Polymeric microelement on the top of the fiber formation and optical loss in this element analysis. Natural Science. 2. 8: 868-872 p.

Chapter 6

# Recent Development in Applications of Cellulose Nanocrystals for Advanced Polymer-Based Nanocomposites by Novel Fabrication Strategies

Chengjun Zhou and Qinglin Wu

Additional information is available at the end of the chapter

http://dx.doi.org/10.5772/46512

## 1. Introduction

Cellulose, one of the world's most abundant, natural and renewable biopolymer resources, is widely present in various forms of biomasses, such as trees, plants, tunicate and bacteria. Cellulose molecule consists of β-1, 4-D-linked glucose chains with molecular formula of $(C_6H_{10}O_5)_n$ (n ranging from 10,000 to 15,000) through an acetal oxygen covalently bonding C1 of one glucose ring and C4 of the adjoining ring (OSullivan, 1997; Samir et al., 2005). In plant cell walls, approximately 36 individual cellulose molecule chains connect with each other through hydrogen bonding to form larger units known as elementary fibrils, which are packed into larger microfibrils with 5-50 nm in diameter and several micrometers in length. These microfibrils have disordered (amorphous) regions and highly ordered (crystalline) regions. In the crystalline regions, cellulose chains are closely packed together by a strong and highly intricate intra- and intermolecular hydrogen-bond network (Figure 1), while the amorphous domains are regularly distributed along the microfibrils. When lignocellulosic biomass are subjected to pure mechanical shearing, and a combination of chemical, mechanical and/or enzymatic treatment (Beck-Candanedo et al., 2005; Bondeson et al., 2006; Filson et al., 2009), the amorphous regions of cellulose microfibrils are selectively hydrolyzed under certain conditions because they are more susceptible to be attacked in contrast to crystalline domains. Consequently, these microfibrils break down into shorter crystalline parts with high crystalline degree, which are generally referred to as cellulose nanocrystals (CNCs) (Habibi et al., 2010). CNCs are also named as microcrystals, whiskers, nanoparticles, microcrystallites, nanofibers, or nanofibrils in the liturautes, all of which are called "cellulose nanocrystals" in this review.

During the past decade, CNCs have attracted considerable attention attributed to their unique features. First, CNCs have nanoscale dimensions and excellent mechanical properties. The theoretical value of Young's modulus along the chain axis for perfect native CNCs is estimated to be 167.5 GPa, which is even theoretically stronger than steel and similar to Kevlar (Tashiro & Kobayashi, 1991), while elastic modulus of native CNCs from cotton and tunicate reach up to 105 and 143 GPa, respectively (Rusli & Eichhorn, 2008; Sturcova et al., 2005). Due to an abundance of hydroxyl groups existed on surface of CNCs, reactive CNCs can be modified with various chemical groups to accomplish expected surface modification, such as esterification, etherification, oxidation, silylation, or polymer grafting, which could successfully functionalize the CNCs and facilitate the incorporation and dispersion of CNCs into different polymer matrices (Habibi et al., 2010). Therefore, CNCs are considered as one of the ideal nano-reinforcements for polymer matrices (including water-soluble and water-insoluble polymer systems) and have already been incorporated into many polymer matrices to produce reinforced composites (Cao et al., 2011; Kvien et al., 2005). In addition, high aspect ratio, low density, low energy consumption, inherent renewability, biodegradability and biocompatibility are also the advantages of environmentally-friendly CNCs (Siro & Plackett, 2010). Because of the growing interest in the bioconversion of renewable lignocellulosic biomass and unsurpassed quintessential physical and chemical properties of CNCs mentioned above, substantial academic and industrial interests have been directed toward the potential applications of CNCs in polymer-based nanocomposites for various fields, such as high performance materials, electronics, catalysis, biomedical, and energy (Duran et al., 2011; Mangalam et al., 2009).

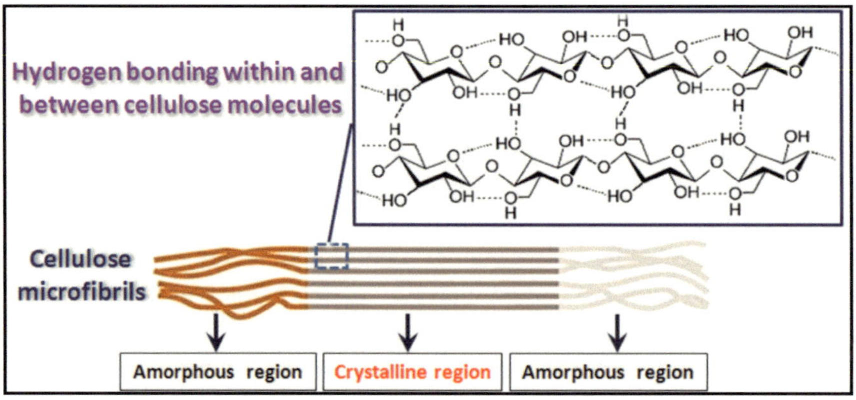


**Figure 1.** Scheme of interaction between cellulose molecular chains within the crystalline region of cellulose microfibrils .

Many different approaches to fabricate polymer/CNCs nanocomposites have been reported (Eichhorn, 2011; Habibi et al., 2010; Moon et al., 2011), and most researches focused on conventional film materials (Peng et al., 2011). Recently, several non-conventional routes of producing polymer/CNCs nanocomposites have been reported, and some of the most exciting developments have been CNC-filled nanocomposite hydrogels (Capadona et al., 2009; Capadona et al., 2008; Shanmuganathan et al., 2010; Zhou et al., 2011b; Zhou et al., 2011c) and electrospun nanofibers (Lu et al., 2009; Martinez-Sanz et al., 2011; Medeiros et al., 2008; Park et al., 2007; Peresin et al., 2010b; Rojas et al., 2009; Xiang et al., 2009; Zhou et al., 2011a; Zoppe et al., 2009). These approaches could help expand novel applications of natural biomass nanocrystals in tissue engineering scaffolds, drug delivery, electronic components and devices.

This review is aim at presenting a summary on recent development of cellulose nanocrystals applied in advanced polymer-based nanocomposites using the novel fabrication strategies for targeting nanocomposite hydrogels and electrospun fibers. Specific attentions will be given to highlight opportunities of above-mentioned nanocomposites for future research.

## 2. Nanocomposite polymer hydrogels

The nanocomposite polymer hydrogels (NPHs), referred to cross-linked polymer networks swollen with water in the presence of nanoparticles or nanostructures, are new generation materials that can be used in a wide variety of applications including stimuli-responsive sensors and actuators, microfluidics, catalysis, separation devices, pharmaceutical, and biomedical devices (Schexnailder & Schmidt, 2009). The most potential use of NPHs is for novel biomaterials in tissue engineering, drug delivery, and hyperthermia treatment because they, in comparison with conventional hydrogels, can provide improved properties such as increased mechanical strength and ability for remote controlling (Samantha A. Meenach, 2009). Because of the excellent dispersion of CNCs in water (Liang et al., 2007), the fabrication, molding, and application of hydrogels containing CNCs without modification have many advantages compared with other nanofillers such as polymer and metal nanoparticles (Saravanan et al., 2007; Wu et al., 2009). Moreover, CNCs possessed the long-term biocompatibility and controlled biodegradability, which is beneficial to further develop applications of NPHs used as biomaterials.

Nakayama et al (2004) for the first time reported cellulose-polymer nanocomposite hydrogels composed of bacterial cellulose (BC) and gelatin. Bacterial cellulose is biosynthesized by microorganisms, and displays unique properties, including high mechanical strength, high water absorption capacity, high crystallinity, and an ultra-fine and highly pure fiber (10-100 nm) network structure (Vandamme et al., 1998). By immersing BC gel in aqueous gelatin solution followed by cross-linking with *N*-(3-dimethylaminopropyl)-*N'*-ethylcarbodiimide hydrochloride, high mechanical strength double-network (DN) nanocomposite hydrogels were prepared. The compressive fracture strength and elastic modulus of the obtained BC-gelatin DN hydrogel are several orders of magnitude higher than those of pure gelation gel, almost equivalent to those of articular cartilage. In addition, this double network nanocomposite hydrogels exhibits not only a mechanical strength as high as several megapascals but also a low frictional coefficient of the order of $10^{-3}$. Buyanov et al (2010) have fabricated high strength composite bacterial cellulose-polyacrylamide (BC–PAM) nanocomposite hydrogels by synthesizeing PAM networks inside BC matrices. These hydrogels not only exhibit superior mechanical properties (compression strength of up to 10 MPa) and withstanding long-term cyclic stresses (up to 2000–6000 cycles) without substantial reduction of mechanical properties, but also show anisotropic behavior on both swelling and deformation. The above-mentioned reports mainly focused on the research used BC as the first nano-network, resulting in its high loading in nanocomposites. However, BC is of the high cost (about 100 times more than that of plant cellulose (Bochek, 2008)) with the relatively low production capacity, likely limiting its potential application in hydrogels for widespread uses. Cellulose nanocrytals or nanofibers isolated from plants have lower cost and higher price-performance ratio than BC, and their size can be facilely adjusted to meet the requirement of hydrogels properties for the various applications.

By employing rod-shaped CNCs (about 10 nm in diameter and 120 nm in length) as the reinforcement nanofiller and in situ free-radical polymerizing/cross-linking acrylamide, we have successfully fabricated PAM-CNC nanocomposite hydrogels (Zhou et al., 2011b). During the gelation reaction of nanocomposite hydrogels, CNCs can accelerate the formation of hydrogels and then increased the effective crosslink density of hydrogels through the grafting copolymerization of monomer acrylamide on the surface of CNCs. Compared to the pure PAM hydrogels, the obtained nanocomposite hydrogels at a low loading level (6.7 wt%) of CNCs exhibited a dramatic enhancement in the shear storage modulus (4.6-fold) and the compression strength (2.5-fold). Hence, CNCs are not only a reinforcing agent for hydrogels, but also act as a multifunctional cross-linker for gelation. A possible mechanism for forming NPHs was proposed, as shown in Figure 2. Moreover, CNCs with smaller dimension and aspect ratios could help promote the sol-gel transition and facilitate the formation of nework of PAM-CNC nanocomposite hydrogels (Zhou et al., 2011c).

Spagnol et al (2012a) synthesized superabsorbent nanocomposite hydrogels based on poly(acrylamide-co-acrylate) (PAM-AA) and CNCs by free-radical aqueous copolymerization, and focused on the investigation of pH-responsiveness and cation-sensitivity character of NPHs. Swelling capacity and swelling-deswelling behavior of PAM-AA/CNC nanocompoiste hydrogel exhibited high pH-sensitivity and reversible pH-responsiveness properties. Furthermore, the swelling measurement in different salt solutions showed that the swelling capacity of NPHs in $MgCl_2$ and $CaCl_2$ solutions are much lower than that in NaCl and KCl solutions and distillated water. The swelling–deswelling process of NPHs was alternatively carried out between sodium and calcium solutions, suggesting a swelling-deswelling pulsatile behavior. It was reported that the pH and salt responsive behavior also occurred in superabsorbent nancomposite hydrogels based on CNCs and chitosan-graft-poly(acrylic acid) copolymer (Spagnol et al., 2012b).

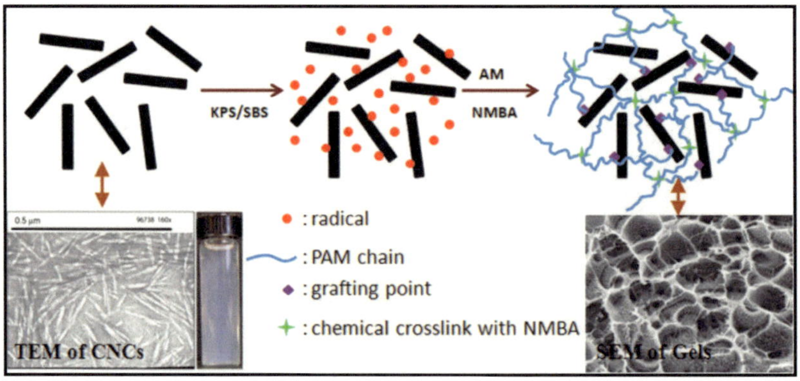


**Figure 2.** Scheme of the gelation mechanisms of PAM–CNC nanocomposite hydrogels. Reprinted with permission from (Zhou et al., 2011b) Copyright 2011 Elsevier.

PAM and its derivative hydrogels have been of great interest for developing applications in tissue engineering because of their non-toxic and biologically inertness, capability for preserving their shape and mechanical strength, and convenient adjustability of mechanical, chemical and biophysical properties (Zhou & Wu, 2011; Zhou et al., 2011d). Recently, we investigated the proliferations of dental follicle stem cells on PAM hydrogel and PAM/CNC nanocomposite hydrogel by alamarBlue® assay. As shown in Figure 3, the cell proliferations on PAM-CNC nanocomposite hydrogels were significantly higher than on the PAM hydrogels at days of 7 and 10, indicating that CNCs could help accelerate the proliferation of stem cells likely because the embedded nanocrystals formed an extracellular matrix-like microstructure in the hydrogel to promote cell-matrix interactions by providing more binding sites for cell adhesion and proliferation. This suggests that CNCs reinforced nanocomposite hydrogel systems showed improved cell biocompatibility and are suitable substrates used as cell carriers and traditional bone-defect repair and bone tissue engineering.

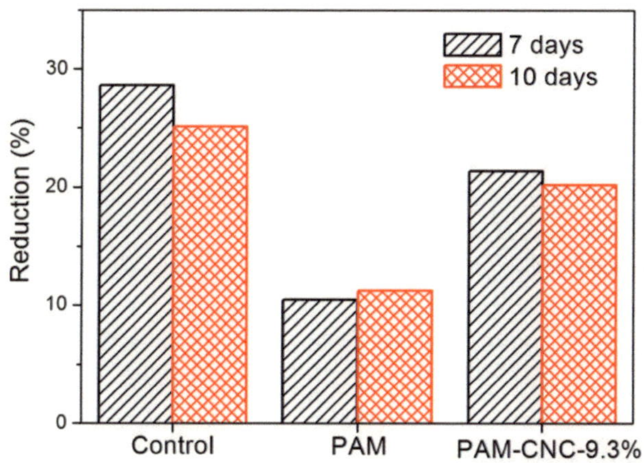


**Figure 3.** Reduction in expression of dental follicle stem cells for pure PAM hydrogels and PAM-CNC nanocomposite hydrogels with 9.3 wt% of CNCs at 7 and 10 days.

Aouada et al (2011) reported a simple, fast, and low cost strategy for the synthesis of nanocomposites by directly immerseing dry polyacrylamide-methylcellulose (PAM-MC) hydrogels into CNC aqueous suspensions. The CNCs were effectively anchored into the hydrogel network to provoke the increase in rigidity of the hydrogel networks and the decrease in pore sizes and the formation of three dimensional well-oriented pores. The incorporation of CNCs improved the crystallinity, and the mechanical and structural network properties of nanocomposite hydrogels without negatively impacting their thermal and hydrophilic properties. The value of the maximum compressive stress increased to 4.4 kPa for nanocomposite hydrogels from 2.1 kPa for pure PAM-MC hydrogels. Because of their biodegradability and biocompatibility, these reinforced nanocomposite hydrogels are promising materials for different technological applications, especially in agricultural applications, such as a carrier vehicle for agrochemical controlled release, such as pesticides (Aouada et al., 2010) and nutrients (Bortolin et al., 2012).

# 3. Electrospun nanocomposite fibers

Electrospinning is a highly versatile technique to generate continuous 1D polymeric fibers with diameters ranging from several micrometers down to 100 nanometers or less through a high voltage charged polymer solutions or melts (Reneker & Yarin, 2008). Electrospun nanofibrous materials possess a variety of interesting characteristics such as small dimension, large specific surface area, wide-range porosity, unique physicochemical property, and excellent flexibility for chemical/physical surface functionalization. Hence, electrospun nanofibrous materials not only are being used in research laboratories but also are increasingly applied in industry (Greiner & Wendorff, 2007). Their application includes, but is not limited to, optoelectronics, sensors, catalysis, filtration, energy-related materials and medicine. Electrospun polymeric nanofibers, however, are not sufficiently strong for many applications because of low molecular chain orientation along the fiber long-axis resulting from low stretching forces in the process of fiber formation (Ayutsede et al., 2006). During the past several years, a large number of studies have been conducted to improve mechanical properties of electrospun polymeric nanofibers. Incorporating nanoparticles into polymer matrices is one technique that has been developed and used as one of the most effective methods for reinforcing electrospun nanofibers (Hou et al., 2005; Lu et al., 2009). As one of the strongest and stiffest natural biopolymers, CNCs have been successfully used as highly effective reinforcing nanofillers for improving mechanical properties of various electrospun polymer matrices, as summarized in Table 1. Moreover, research effort has been focused on increased dispersion of CNCs in the matrix, improved alignment of CNCs along the fiber length, tailored CNC-matrix interfacial properties (Moon et al., 2011). According to the origination of polymer (i.e. synthetic and natural polymer), this section will concentrate on the processing of electrospun nanocomposite fibers and the effect of CNCs on their mechanical properties.

## 3.1. CNCs reinforced synthetic polymer

Due to the excellent dispersion property of CNCs in water, the first report of electrospun nanocomposite fibers was to electrospin water-soluble PEO and CNCs from BC. Highly crystalline rod-like CNCs with a high aspect ratio and specific area (420 ± 190 nm in length and 11 ± 4 nm in width) were incorporated into the electrospun PEO fibers with a diameter of less than 1 μm. Well-embedded CNCs were aligned and partially clustered inside the fibers. Compared with the electrospun PEO fibrous mats, the tensile modulus, tensile strength and elongation of electrospun nanocomposite fibrous mats containing 0.4 wt% of CNCs were increased by 193.9%, 72.3% and 233.3%, respectively, indicating that the existence of CNCs effectively improved the mechanical properties of the electrospun mats.

The dispersion of CNCs in electrospun fibers could be improved by tailoring the geometrical dimensions (length, $L$, and width, $w$) of CNCs, which can be controlled by adjusting the source of the cellulosic material and the conditions of fabrication. Recently, wood-based CNCs with a diameter of 10 ± 3 nm and a length of 112 ± 26 nm were processed into electrospun PEO fibers (Zhou et al., 2011a). Figure 4 shows that rod-shaped CNCs without obvious aggregation are well-dispersed in the as-spun nanofibers. Decreasing aspect ratio can facilitate a better dispersion of CNCs within polymer matrix. When the CNC content was increased from 0 to 20 wt %, $E_{max}$ and $\sigma$ of nanocomposite fibrous mats increased from 15.2 to 38.3 MPa and from 2.50 to 7.01 MPa, respectively, whereas $\varepsilon_b$ decreased markedly from 200 to 106 %. In addition, improving dispersion of CNCs within the fibers could also be achieved by adopting some pre-treatment and processing methods on electrospinning solution, such as sonication (Martinez-Sanz et al., 2011).

| Matrix | CNC origin and size | $\phi_{CNC}$ wt% | $E_{max}$ %Increase | $\sigma$ %Increase | $\varepsilon_b$ %Increase | Reference |
|---|---|---|---|---|---|---|
| PEO | Bacteria, 11 ± 4 nm in *w*, 420 ± 190 nm in *L* | 0–0.4 | +193.9 | + 72.3 | +233.3 | (Park et al., 2007) |
| PAA | Cotton | 0–20 | + 3441.1<br>+ 7633.9[a] | +1455.2<br>+ 5658.6[a] | - 73.5<br>- 99.9[a] | (Lu & Hsieh, 2009) |
| PEO | MCC, 10 ± 3 nm in *w*, 112 ± 26 nm in *L* | 0–20 | + 152.0<br>+ 392.1[b] | + 180.4<br>+ 240.8[b] | - 47.0<br>- 37.5[b] | (Zhou et al., 2011a) |
| PEO | Cotton, 5–10 nm in *w*, 40–100 nm in *L* | 0–20 | + 190.5 | + 377.5 | - 33.5 | (Zhou et al., 2012b) |
| PEO | Tunicate, ~ 20 nm in *w*, ~ 2 µm in *L* | 0–15 | + 98.8[c] | — | — | (Changsarn et al., 2011) |
| PVA | Ramie, 3–10 nm in *w*, 100–250 nm in *L* | 0–15 | + 270.9[c] | — | — | (Peresin et al., 2010b) |
| PCL | Ramie, 3–10 nm in *w*, 100–250 nm in *L* | 0–7.5 | + 64.3<br>+ ~40[c] | + 37.2<br>— | + 49.0<br>— | (Zoppe et al., 2009) |
| PS | Paper, 10–20 nm in *w*, 200 nm in *L* | 0–9 | + ~60[c] | — | — | (Rojas et al., 2009) |
| PMMA | Bacteria, 15–20 nm in *w*, 0.3–8 µm in *L* | 0–20 | — | — | — | (Olsson et al., 2010) |
| PMMA | Wood, ~17 nm in *w*, 190–660 nm in *L* | 0–41 | + 17[d] | — | — | (Dong et al., 2012) |
| PLA | MCC, 92 ± 3 nm in *w*, 124 ± 35 nm in *L* | 0–10 | + 37.0 | + 30.2 | -1.4 | (Xiang et al., 2009) |
| EVOH | Bacteria, lower than 30 nm in *w* | 0–8 | — | — | — | (Martinez-Sanz et al., 2011) |
| PLA | Cotton | 0–12.5 | — | +161.9 | — | (Ramirez, 2010) |
| Silk | Bark, 25–40 nm in *w*, 400–500 nm in *L* | 0–4 | + 300.3 | + 208.0 | - 55.6 | (Huang et al., 2011) |

[a], tensile properties of crosslinked nanocomposite fibrous mats; [b], tensile properties of heterogeneous nanocomposite fibrous mats; [c], Dynamic mechanical analysis (DMA) storage modulus of nanocomposite fibrous mats; [d], nano-DMA storage modulus of individual nanocomposite fiber.

*w*, width of CNCs; *L*, length of CNCs; $\phi_{CNC}$, loading range of CNCs in composites; $E_{max}$, max Young's modulus of nanocomposite mats; $\sigma$, max tensile stress at yield for nanocomposite mats with $E_{max}$; $\varepsilon_b$, elongation at break for nanocomposite mats with $E_{max}$.

Acronyms: PEO, poly(ethylene oxide); PVA, poly(vinyl alcohol); PAA, poly(acrylic acid); PCL, poly(ε-caprolactone); PLA, poly(lactic acid); PS, polystyrene; EVOH, Ethylene–vinyl alcohol copolymer; PMMA, poly(methyl methacrylate);

**Table 1.** Summary of the experimental results of the reviewed publications involving origin and size of CNCs, and mechanical properties of electrospun nanocomposite fibers/mats.

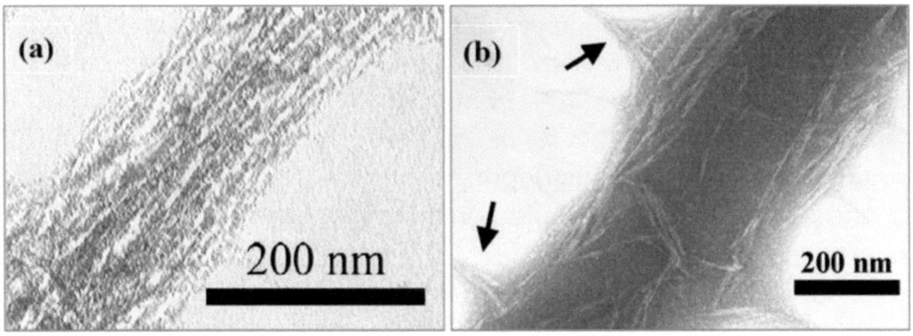


**Figure 4.** TEM micrographs of PEO/CNC nanocomposite fibers with 20 wt% CNC loading electrospun from 5 wt% (a) and 7 wt% (b) solutions. (arrows pointing to typical secondary nanofibers). Reprinted with permission from (Zhou et al., 2011a) Copyright 2011 American Chemical Society.

For the hydrophobic polymer systems, CNCs should firstly be well-dispersed in organic solvent (such as DMF, N,N'-dimethylformamide; THF, tetrahydrofuran) before electrospinning because solution-based processing of nanocomposite nanofibers require good dispersability of both CNCs and polymer in one common solvent. Direct ultrasonication of freeze-dried CNCs is a facile method to prepare electrospinning solution (Xiang et al., 2009), and the redispersion of CNCs could be improved by conducting surface modification on CNCs (Zoppe et al., 2009) and adding additves. Rojas et al (2009) obtained fibers with smooth surface using solely THF as a solvent to electrospin PS containing surfactant-dispersed CNCs. The addition of non-ionic surfactant to the PS-CNC suspensions improved their stability, minimized (or prevented) the presence of beads in the resulting fibers, and promoted the nano-reinforcement of CNCs on PS fibers. However, the directly re-dispersed CNCs cannot form stable suspension at high loading leverls because the hydrophilic nature of cellulose and the strong hydrogen bonding interactions between CNCs. Olsson et al (2010) reported a two-step solvent exchange method to replace the water of CNC suspension by acetone, followed by further replacement through DMF/THF solvent in the same manner. A high degree of dispersion of CNCs was obtained for a variety of CNC contents and the aggregation of CNCs up to 7 wt% was greatly suppressed because CNCs were aligned and rapidly sealed inside PMMA matrix during the continuous formation of electrospun fibers. Moreover, the direct solvent exchange from water to organic solvent conducted by vacuum rotary evaporation was also used to disperse CNCs (Dong et al., 2012; Zoppe et al., 2009) .

The alignment of CNCs along the fiber is also an important factor to determine the axial strength of electrospun nanocomposite fibers reinforced with CNCs. Considering CNCs can be aligned under the high electrostatic fields (Habibi et al., 2008), the electrospinning process could facilitate alignment of CNCs along the fiber long-axis. Usually, the alignment of CNCs was observed by scanning electron microscope (SEM) (Dong et al., 2012; Olsson et al., 2010) and TEM (Changsarn et al., 2011; Lu & Hsieh, 2009; Park et al., 2007; Zhou et al., 2011a). Figure 4 presents typical TEM pictures of CNC aligned parallel inside/along the longitudal axis of nanocomposite fibers. Interestingly, Figure 4b shows that CNCs dispensed in electrospun fibers had radial anisotropy or a skin-core morphology, in which CNCs in the core are oriented more randomly, while ones in the skin have a higher degree of orientation. Dong et al (2012) investigated the orientation of CNCs embedded in electrospun PMMA/CNC fiber by solvent-etching PMMA from nanocomposite fibers with drops of THF, and discovered highly aligned CNCs along the fiber axis, as shown in Figure 5. Besides the high electrostatic fields, the high alignment of CNCs along the polymer fiber could also be attributed to the shear forces in the liquid jet and the orientation of polymer chains during electrospinning process, and the nanoscale confinement effect (Chen et al., 2009).

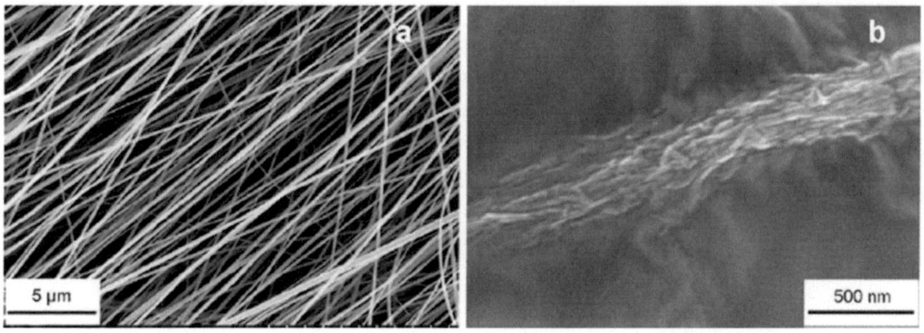


**Figure 5.** SEM of (a) alignment of PMMA/CNC fibers with 33 wt% of CNCs and (b) alignment of CNCs along fiber long-axis revealed by solvent etching. Reprinted with permission from (Dong et al., 2012) Copyright 2012 Elsevier.

It is well known that the improved interface between nanofillers with polymer matrix is beneficial to the mechanical properties of polymer-based nanocomposites (Zhang et al., 2010; Zhou et al., 2012a; Zhou et al., 2008b; Zhou et al., 2008c). For hydrophilic polymers, the hydrogen bonding between CNCs and polymer matrix play a very important role in determining polymer-CNC interaction. Peresin et al (Peresin et al., 2010b) reported CNC-reinforced nanocomposite fibers produced via electrospinning of poly(vinyl alcohol) (PVA) with two different concentrations of acetyl groups. The hydrogen bonding between PVA and CNCs was confirmed by observing the band between 3550 and 3200 $cm^{-1}$ in Fourier transform infrared spectra (FTIR). The higher the hydrolysis degree of PVA (i. e. more –OH group in PVA chain), the stronger the PVA-CNC interaction, which was also observed by FTIR spectra. To confirm the effect of the hydrogen bonding, DMA in tensile mode was used for mechanical analysis. The storage modulus of nanocomposite fibrous mats showed a steady increase with increased CNC content from 0 to 15% loading. The storage modulus was 15.45 MPa for pure PVA mats and rose to 57.30 MPa at 15 wt% CNC loading. It was concluded that the observed strength enhancement in CNC-loaded PVA mats mainly is related to the reinforcing effect of the dispersed CNCs through the percolation network held by hydrogen bonds.

In addition to strong hydrogen bonding, covalent bonding also provides a means for enhancing polymer-nanofiller interface to achieve optimal composite properties (Zhou et al., 2008a). To produce greater reinforcing effect from CNCs in electrospun nanocomposite fibers, Lu & Hsieh (2009) fabricated electrospun PAA-CNC nanocomposite fibers. The interfacial interactions between CNCs and PAA could be further improved by heat-induced esterification between the CNC surface hydroxyls and PAA carboxyl groups, which produce covalent crosslinks at the PAA-CNC interfaces, render the nanocomposite fibrous mats insoluble in water, and make mats to be more thermally stable and far more superior tensile properties. The Young's modulus and tensile strength of mats were significantly improved with increased CNC loadings in the nanocomposite fibers by up to 35-fold and 16-fold, respectively, with 20 wt% CNC loading, as shown in Figure 6. It is more impressive that the crosslinked nanocomposite fibrous mats with 20 wt% CNC exhibited 77-fold increase in modulus and 58-fold increase in strength, respectively. Moreover, the synergies of polymer crosslinking network and CNC's reinforcement could also improve the mechanical properties of nanocomposite fibrous mats. Recently, we reported UV-initiated crosslinking of PEO nanofibers in the presence of CNCs, which was performed with pentaerythritol triacrylate as both photo-initiator and crosslinker (Zhou et al., 2012b). With increased CNC content up to 10 wt%, the maximum tensile stress and Young's modulus of the crosslinked PEO/CNC composite fibrous mats increased by 377.5 and 190.5% than those of uncrosslinked PEO mats, and 76.5 and 127.4% than those of crosslinked PEO mats, respectively.

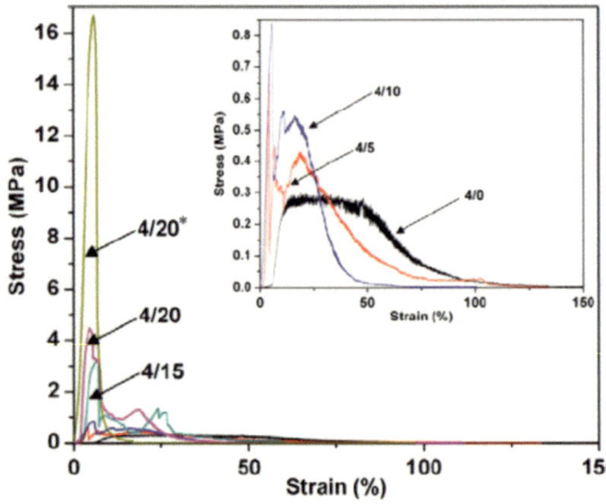


**Figure 6.** Stress-strain curves of electrospun PAA/CNC nanocomposite fibrous mats at different CNC loadings. * represents crosslinked samples. Reprinted with permission from (Lu & Hsieh, 2009) Copyright 2009 IOP Publishing.

Due to the fact that the electrospun mats are non-woven fabrics, their mechanical properties are influenced by several factors including composition, morphology and structure of individual fiber, the interaction between fibers, orientation of fibers, and porosity of mats. An nano-indentation study was performed on single sub-micron PMMA/CNC fibers in the transverse direction to explore the reinforcement of CNCs on single nanocomposite fibers, which showed a modest increase in the mechanical properties with increasing CNC content, about 17% improvement in nano-DMA storage modulus with the loading of 17 wt% CNCs (Dong et al., 2012). In addition to the reinforceing effect of CNCs on fibers, it was also found that the addition of CNCs could reduce electrospun fiber diameters and improved fiber uniformity attributed to the enhanced electric conductivity of electrospinning solutions in the presence of CNCs. This tends to increase the mechanical properties of mats because smaller fiber diameters yield higher overall relative bonded areas between fibers by increasing its surface area, bonding density, and distribution of bonds (Olsson et al., 2010; Peresin et al., 2010b; Zhou et al., 2011a). Moreover, the orientation of nanocomposite fibers within mats also influence greatly on the mechanical properties of mats (Wang et al., 2011). Alignment of electrospun nanocomposite fibers reinforced with CNCs has been achieved by many techniques, including by running the fibers over a hollow spool with high rotation (Olsson et al., 2010), an aluminium frame with openings (Changsarn et al., 2011), a rotating mandrel covered with aluminium foil (Dong et al., 2012). Figure 5a shows the morphology of one-dimensional aligned fibers of PMMA/CNC. Furthermore, the more sufficient contact and stronger bonding between fibers could also lead to the improvement in mechanical properties of fibrous mats. It was found that humidity treatments on the PVA/CNC nanocomposite fibrous mats induced significant enhancement of strength as a result of the increased contact area and enhanced adhesion between the fibers (Peresin et al., 2010a). The heterogeneous fibrous mats composed of rigid–flexible bimodal PEO/CNC nanocomposite fibers was demonstrated to be higher in mechanical properties than their homogeneous counterparts (Zhou et al., 2011a). The reinforced mechanism was illustrated by morphology observation of the tensile process, as shown in Figure 7. With increased strain, the tensile stress of unaligned electrospun fibrous mats increased sharply at the beginning, and then increased slowly over a relatively long period of strain followed by the final rupture (Gomez-Tejedor et al., 2011). At the beginning of tensile process, the mats were stretched in a macroscopic view and most fibers in mats hardly moved attributed to the cohesion between fibers, which could determine Young's modulus of mats. When most fibers in mats reached the tightened form, yield of mats appeared, at which the interaction points among fibers were broken. With further increase of tensile strain, the fibers in mats were drawn out to highly align along the tensile direction. However, the rupture of individual fibers one by one did not influence the tensile properties of the whole mats, resulting in a large elongation at break for electrospun mats. At the maximum tensile stress level, most fibers were necked and broken, leading to the final rupture of mats.

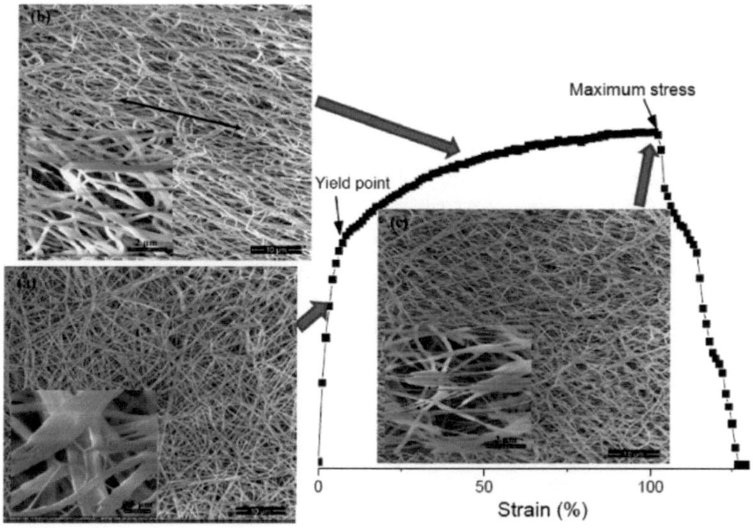


**Figure 7.** Morphology observation of the tensile process on PEO/CNC nanocomposite fibrous mats with 20 wt% CNC loading electrospun from 7 wt% solutions. Reprinted with permission from (Zhou et al., 2011a) Copyright 2011 American Chemical Society

### 3.2. CNCs reinforced natural polymer

With increasing environmental consciousness about petroleum-based polymer materials, development of fully biodegradable, eco-friendly, and sustainable, bio-based nanocomposites have attracted more and more attention both in the academic and industrial fields (Kim & Netravali, 2010; Oksman et al., 2006). The so-called "green" composites are derived from natural resources including plant or animal origin (Khalil et al., 2012). Some researchers have successfully used CNCs as highly effective reinforcing nanofillers to fabricate electrospun bio-nanocomposite fibers from various biopolymers such as PLA (Li, 2010; Ramirez, 2010; Xiang et al., 2009), cellulose and its derivative (Herrera Vargas, 2010; Magalhaes et al., 2009), silk (Huang et al., 2011), and lignin (Ago et al., 2012).

PLA is such typical bio-based aliphatic polyester produced by polymerization of lactic acid, which is originated from renewable natural resources such as corn, starch, and molasses. The electrospun PLA/CNC nanocomposite fibers were widely reported because PLA possessed excellent physical properties of transparency, high elastic modulus, and high melting temperature. Xiang el al (2009) incorporated CNCs into electrospun PLA fibers, and found that Young's modulus and strength of obtained nanocomposite mats with 1 wt% loading of CNCs were improved by approximately 37 and 30 %, respectively. Besides the reinforcement of CNCs on PLA fibers, Ramírez (2010) also investigated the cytocompatibility of the PLA/CNC nanocomposite fibrous mats used as scaffold. After one week of cell culture, confocal microscopy indicated that the cells grown on the PLA/CNC nanocomposite mats were confluent and very well aligned along the fibers while cells cultured on pure PLA mats were not as confluent as in the developed nanocomposite mats. This demonstrates the feasibility of the PLA/CNC nanocomposite fibrous mats as a potential scaffold for bone tissue engineering.

Magalhães et al (2009) reported a electrospun fully-cellulosic core-in-shell nanocomposite fibers consisting of regenerated cellulose (type II and amorphous) in the shell and CNCs in the core, which were fabricated by the co-electrospinning technique. Wood-based cellulose was dissolved in *N*-methyl morpholine oxide at 120 °C and diluted with dimethyl sulphoxide, and used in an external concentric capillary needle as the sheath (shell) solution. At the same time, a CNC suspension obtained by the sulphuric acid hydrolysis of sisal bleached and cotton fibers was used as the core liquid in the internal concentric capillary needle. It was found that the formation of individual fiber could be promoted by precisely controlling the voltage and flow rate to decrease the shell-to-core volume ratio. The novel core-in-shell nanocomposite fibers also showed better mechanical properties than the pure electrospun cellulose II fibers.

## 4. Conclusion and perspectives for the future

In the review, recent development on applications of cellulose nanocrystals in nanocomposites fabricated by two novel strategies, i.e., gelation and electrospinning is presented. It is shown that CNCs have a distinct advantage for improving mechanical properties of both nanocomposite hydrogels and electrospun nanocomposite fibers/mats.

The obviously reinforced effect of CNCs could help facilitate the potential applications of CNC-filled nanocompsite as advanced function materials.

CNCs reinforced hydrogels have been reported widely, and their properties, mainly mechanical properties, have been investigated. With the increased requirement for the multifunctional properties, nanocomposite hydrogels from CNCs and other stimuli-responsive polymer would be further developed. Various nanocomposite hydrogels reinforced with CNCs can be designed to become fast temperature, pH, and salt sensitive for controllable drug delivery system. Futhermore, using CNCs to reinforce natural polymer-based hydrogels could endow many favorable properties such as hydrophilicity, biodegradability, biocompatibility, low cost, and non-toxicity, resulting in applications of nanocomposite hydrogels in tissue engineering.

For the electrospun nanocomposite fibers containing CNCs, most studies focus on their fabrication, morphology, mechanical and thermal properties. There are still several major challenges for the further development of CNC-reinforced nanocomposites fibers. These include surface modification and homogeneous dispersion of CNCs, interface and alignment characterization of CNCs within individual electrospun nanocomposite fiber, analytical model for mechanics of single nanocomposite fiber, and assembly and effect of nanocomposite fibers within mats. More importantly, it is very worthwhile to exploit the functional characteristics and properties of CNC-filled nanocompsite fibers/mats to create new and specific applications such as energy-related materials, sensor, barrier films, and tissue engineering scaffolds.

## Author details

Chengjun Zhou and Qinglin Wu
*School of Renewable Natural Resource, Louisiana State University Agricultural Center, Baton Rouge, Louisiana, USA*

## Acknowledgement

The authors would like to acknowledge the financial support from Louisiana Board of Regents Industrial Tie Subprogram (LEQSF(2010-13)-RD-B-01) and the USDA Rural Development Biomass Initiative Program (68-3A75-6-508). Special thanks go to Mrs. Maryam Rezai Rad and Dr. Shaomian Yao for their assistance in the cell culture of PAM/CNC nanocomposite hydrogels.

## 5. References

[1] Ago, M., Okajima, K., Jakes, J. E., Park, S., & Rojas, O. J. (2012). Lignin-Based Electrospun Nanofibers Reinforced with Cellulose Nanocrystals. *Biomacromolecules,* Vol.13, No.3, pp. 918-926.

[2] Aouada, F. A., de Moura, M. R., Orts, W. J., & Mattoso, L. H. C. (2010). Polyacrylamide and methylcellulose hydrogel as delivery vehicle for the controlled release of paraquat pesticide. *Journal of Materials Science,* Vol.45, No.18, pp. 4977-4985.

[3] Aouada, F. A., de Moura, M. R., Orts, W. J., & Mattoso, L. H. C. (2011). Preparation and Characterization of Novel Micro- and Nanocomposite Hydrogels Containing Cellulosic Fibrils. *Journal of Agricultural and Food Chemistry,* Vol.59, No.17, pp. 9433-9442.

[4] Ayutsede, J., Gandhi, M., Sukigara, S., Ye, H. H., Hsu, C. M., Gogotsi, Y., & Ko, F. (2006). Carbon nanotube reinforced Bombyx mori silk nanofibers by the electrospinning process. *Biomacromolecules,* Vol.7, No.1, pp. 208-214.

[5] Beck-Candanedo, S., Roman, M., & Gray, D. G. (2005). Effect of reaction conditions on the properties and behavior of wood cellulose nanocrystal suspensions. *Biomacromolecules,* Vol.6, No.2, pp. 1048-1054.

[6] Bochek, A. M. (2008). Prospects for use of polysaccharides of different origin and environmental problems in processing them. *Fibre Chemistry,* Vol.40, No.3, pp. 192-197.

[7] Bondeson, D., Mathew, A., & Oksman, K. (2006). Optimization of the isolation of nanocrystals from microcrystalline cellulose by acid hydrolysis. *Cellulose,* Vol.13, No.2, pp. 171-180.

[8] Bortolin, A., Aouada, F. A., de Moura, M. R., Ribeiro, C., Longo, E., & Mattoso, L. H. C. (2012). Application of Polysaccharide Hydrogels in Adsorption and Controlled-Extended Release of Fertilizers Processes. *Journal of Applied Polymer Science,* Vol.123, No.4, pp. 2291-2298.

[9] Buyanov, A. L., Gofman, I. V., Revel'skaya, L. G., Khripunov, A. K., & Tkachenko, A. A. (2010). Anisotropic swelling and mechanical behavior of composite bacterial cellulose-poly(acrylamide or acrylamide-sodium acrylate) hydrogels. *Journal of the Mechanical Behavior of Biomedical Materials,* Vol.3, No.1, pp. 102-111.

[10] Cao, X. D., Habibi, Y., Magalhaes, W. L. E., Rojas, O. J., & Lucia, L. A. (2011). Cellulose nanocrystals-based nanocomposites: fruits of a novel biomass research and teaching platform. *Current Science,* Vol.100, No.8, pp. 1172-1176.

[11] Capadona, J. R., Shanmuganathan, K., Triftschuh, S., Seidel, S., Rowan, S. J., & Weder, C. (2009). Polymer Nanocomposites with Nanowhiskers Isolated from Microcrystalline Cellulose. *Biomacromolecules,* Vol.10, No.4, pp. 712-716.

[12] Capadona, J. R., Shanmuganathan, K., Tyler, D. J., Rowan, S. J., & Weder, C. (2008). Stimuli-responsive polymer nanocomposites inspired by the sea cucumber dermis. *Science,* Vol.319, No.5868, pp. 1370-1374.

[13] Changsarn, S., Mendez, J. D., Shanmuganathan, K., Foster, E. J., Weder, C., & Supaphol, P. (2011). Biologically Inspired Hierarchical Design of Nanocomposites Based on Poly(ethylene oxide) and Cellulose Nanofibers. *Macromolecular Rapid Communications,* Vol.32, No.17, pp. 1367-1372.

[14] Chen, D., Liu, T. X., Zhou, X. P., Tjiu, W. C., & Hou, H. Q. (2009). Electrospinning Fabrication of High Strength and Toughness Polyimide Nanofiber Membranes Containing Multiwalled Carbon Nanotubes. *Journal of Physical Chemistry B,* Vol.113, No.29, pp. 9741-9748.

[15] Dong, H., Strawhecker, K. E., Snyder, J. F., Orlicki, J. A., Reiner, R. S., & Rudie, A. W. (2012). Cellulose nanocrystals as a reinforcing material for electrospun poly(methyl

methacrylate) fibers: Formation, properties and nanomechanical characterization. *Carbohydrate Polymers,* Vol.87, No.4, pp. 2488-2495.

[16] Duran, N., Lemes, A. P., Duran, M., Freer, J., & Baeza, J. (2011). A Minireview of Cellulose Nanocrystals and Its Potential Integration as Co-Product in Bioethanol Production. *Journal of the Chilean Chemical Society,* Vol.56, No.2, pp. 672-677.

[17] Eichhorn, S. J. (2011). Cellulose nanowhiskers: promising materials for advanced applications. *Soft Matter,* Vol.7, No.2, pp. 303-315.

[18] Filson, P. B., Dawson-Andoh, B. E., & Schwegler-Berry, D. (2009). Enzymatic-mediated production of cellulose nanocrystals from recycled pulp. *Green Chemistry,* Vol.11, No.11, pp. 1808-1814.

[19] Gomez-Tejedor, J. A., Van Overberghe, N., Rico, P., & Ribelles, J. L. G. (2011). Assessment of the parameters influencing the fiber characteristics of electrospun poly(ethyl methacrylate) membranes. *European Polymer Journal,* Vol.47, No.2, pp. 119-129.

[20] Greiner, A., & Wendorff, J. H. (2007). Electrospinning: A fascinating method for the preparation of ultrathin fibres. *Angewandte Chemie-International Edition,* Vol.46, No.30, pp. 5670-5703.

[21] Habibi, Y., Heim, T., & Douillard, R. (2008). AC electric field-assisted assembly and alignment of cellulose nanocrystals. *Journal of Polymer Science Part B-Polymer Physics,* Vol.46, No.14, pp. 1430-1436.

[22] Habibi, Y., Lucia, L. A., & Rojas, O. J. (2010). Cellulose Nanocrystals: Chemistry, Self-Assembly, and Applications. *Chemical Reviews,* Vol.110, No.6, pp. 3479-3500.

[23] Herrera Vargas, N. (2010). Aligned cellulose nanofibers prepared by electrospinning. *Applied Physics and Mechanical Engineering* (Vol. Master, p. 58). Luleå: Luleå University of Technology.

[24] Hou, H. Q., Ge, J. J., Zeng, J., Li, Q., Reneker, D. H., Greiner, A., & Cheng, S. Z. D. (2005). Electrospun polyacrylonitrile nanofibers containing a high concentration of well-aligned multiwall carbon nanotubes. *Chemistry of Materials,* Vol.17, No.5, pp. 967-973.

[25] Huang, J., Liu, L., & Yao, J. M. (2011). Electrospinning of Bombyx mori Silk Fibroin Nanofiber Mats Reinforced by Cellulose Nanowhiskers. *Fibers and Polymers,* Vol.12, No.8, pp. 1002-1006.

[26]Khalil, H. P. S. A., Bhat, A. H., & Yusra, A. F. I. (2012). Green composites from sustainable cellulose nanofibrils: A review. *Carbohydrate Polymers,* Vol.87, No.2, pp. 963-979.

[27] Kim, J. T., & Netravali, A. N. (2010). Mechanical, Thermal, and Interfacial Properties of Green Composites with Ramie Fiber and Soy Resins. *Journal of Agricultural and Food Chemistry,* Vol.58, No.9, pp. 5400-5407.

[28] Kvien, I., Tanem, B. S., & Oksman, K. (2005). Characterization of cellulose whiskers and their nanocomposites by atomic force and electron microscopy. *Biomacromolecules,* Vol.6, No.6, pp. 3160-3165.

[29] Li, Y. (2010). Emulsion electrospinning of nanocrystalline cellulose reinforced nanocomposite fibres. *Materials Engineering* (Vol. Master of Applied Science, p. 115). Vancouver: University of British Columbia.

[30] Liang, S. M., Zhang, L. N., Li, Y. F., & Xu, J. (2007). Fabrication and properties of cellulose hydrated membrane with unique structure. *Macromolecular Chemistry and Physics,* Vol.208, No.6, pp. 594-602.

[31] Lu, P., & Hsieh, Y. L. (2009). Cellulose nanocrystal-filled poly(acrylic acid) nanocomposite fibrous membranes. *Nanotechnology,* Vol.20, No.41, pp. 415604.

[32] Lu, X. F., Wang, C., & Wei, Y. (2009). One-Dimensional Composite Nanomaterials: Synthesis by Electrospinning and Their Applications. *Small,* Vol.5, No.21, pp. 2349-2370.

[33] Magalhaes, W. L. E., Cao, X. D., & Lucia, L. A. (2009). Cellulose Nanocrystals/Cellulose Core-in-Shell Nanocomposite Assemblies. *Langmuir,* Vol.25, No.22, pp. 13250-13257.

[34] Mangalam, A. P., Simonsen, J., & Benight, A. S. (2009). Cellulose/DNA Hybrid Nanomaterials. *Biomacromolecules,* Vol.10, No.3, pp. 497-504.

[35] Martinez-Sanz, M., Olsson, R. T., Lopez-Rubio, A., & Lagaron, J. M. (2011). Development of electrospun EVOH fibres reinforced with bacterial cellulose nanowhiskers. Part I: Characterization and method optimization. *Cellulose,* Vol.18, No.2, pp. 335-347.

[36] Medeiros, E. S., Mattoso, L. H. C., Ito, E. N., Gregorski, K. S., Robertson, G. H., Offeman, R. D., Wood, D. F., Orts, W. J., & Imam, S. H. (2008). Electrospun Nanofibers of Poly(vinyl alcohol) Reinforced with Cellulose Nanofibrils. *Journal of Biobased Materials and Bioenergy,* Vol.2, No.3, pp. 231-242.

[37] Moon, R. J., Martini, A., Nairn, J., Simonsen, J., & Youngblood, J. (2011). Cellulose nanomaterials review: structure, properties and nanocomposites. *Chemical Society Reviews,* Vol.40, No.7, pp. 3941-3994.

[38] Nakayama, A., Kakugo, A., Gong, J. P., Osada, Y., Takai, M., Erata, T., & Kawano, S. (2004). High mechanical strength double-network hydrogel with bacterial cellulose. *Advanced Functional Materials,* Vol.14, No.11, pp. 1124-1128.

[39] Oksman, K., Mathew, A. P., Bondeson, D., & Kvien, I. (2006). Manufacturing process of cellulose whiskers/polylactic acid nanocomposites. *Composites Science and Technology,* Vol.66, No.15, pp. 2776-2784.

[40] Olsson, R. T., Kraemer, R., Lopez-Rubio, A., Torres-Giner, S., Ocio, M. J., & Lagaron, J. M. (2010). Extraction of Microfibrils from Bacterial Cellulose Networks for Electrospinning of Anisotropic Biohybrid Fiber Yarns. *Macromolecules,* Vol.43, No.9, pp. 4201-4209.

[41] OSullivan, A. C. (1997). Cellulose: the structure slowly unravels. *Cellulose,* Vol.4, No.3, pp. 173-207.

[42] Park, W.-I., Kang, M., Kim, H.-S., & Jin, H.-J. (2007). Electrospinning of Poly(ethylene oxide) with Bacterial Cellulose Whiskers. *Macromolecular Symposia,* Vol.249-250, No.1, pp. 289-294.

[43] Peng, B. L., Dhar, N., Liu, H. L., & Tam, K. C. (2011). Chemistry and Applications of Nanocrystalline Cellulose and Its Derivatives: A Nanotechnology Perspective. *Canadian Journal of Chemical Engineering,* Vol.89, No.5, pp. 1191-1206.

[44] Peresin, M. S., Habibi, Y., Vesterinen, A. H., Rojas, O. J., Pawlak, J. J., & Seppala, J. V. (2010a). Effect of Moisture on Electrospun Nanofiber Composites of Poly(vinyl alcohol) and Cellulose Nanocrystals. *Biomacromolecules,* Vol.11, No.9, pp. 2471-2477.

[45] Peresin, M. S., Habibi, Y., Zoppe, J. O., Pawlak, J. J., & Rojas, O. J. (2010b). Nanofiber Composites of Polyvinyl Alcohol and Cellulose Nanocrystals: Manufacture and Characterization. *Biomacromolecules,* Vol.11, No.3, pp. 674-681.

[46] Ramirez, M. A. (2010). Cellulose Nanocrystals Reinforced Electrospun Poly(lactic acid) Fibers as Potential Scaffold for Bone Tissue Engineering. *Department of Wood & Paper Science* (Vol. Master of Science, p. 75). Raleigh: North Carolina State University.

[47] Reneker, D. H., & Yarin, A. L. (2008). Electrospinning jets and polymer nanofibers. *Polymer,* Vol.49, No.10, pp. 2387-2425.

[48] Rojas, O. J., Montero, G. A., & Habibi, Y. (2009). Electrospun Nanocomposites from Polystyrene Loaded with Cellulose Nanowhiskers. *Journal of Applied Polymer Science,* Vol.113, No.2, pp. 927-935.

[49] Rusli, R., & Eichhorn, S. J. (2008). Determination of the stiffness of cellulose nanowhiskers and the fiber-matrix interface in a nanocomposite using Raman spectroscopy. *Applied Physics Letters,* Vol.93, No.3, pp.

[50] Samantha A. Meenach, K. W. A., and J. Zach Hilt. (2009). *Hydrogel Nanocomposites: Biomedical Applications, Biocompatibility, and Toxicity Analysis*. New York: Springer.

[51] Samir, M. A. S. A., Alloin, F., & Dufresne, A. (2005). Review of recent research into cellulosic whiskers, their properties and their application in nanocomposite field. *Biomacromolecules,* Vol.6, No.2, pp. 612-626.

[52] Saravanan, P., Raju, M. P., & Alam, S. (2007). A study on synthesis and properties of Ag nanoparticles immobilized polyacrylamide hydrogel composites. *Materials Chemistry and Physics,* Vol.103, No.2-3, pp. 278-282.

[53] Schexnailder, P., & Schmidt, G. (2009). Nanocomposite polymer hydrogels. *Colloid and Polymer Science,* Vol.287, No.1, pp. 1-11.

[54] Shanmuganathan, K., Capadona, J. R., Rowan, S. J., & Weder, C. (2010). Bio-inspired mechanically-adaptive nanocomposites derived from cotton cellulose whiskers. *Journal of Materials Chemistry,* Vol.20, No.1, pp. 180-186.

[55] Siro, I., & Plackett, D. (2010). Microfibrillated cellulose and new nanocomposite materials: a review. *Cellulose,* Vol.17, No.3, pp. 459-494.

[56] Spagnol, C., Rodrigues, F. H. A., Neto, A. G. V. C., Pereira, A. G. B., Fajardo, A. R., Radovanovic, E., Rubira, A. F., & Muniz, E. C. (2012a). Nanocomposites based on poly(acrylamide-co-acrylate) and cellulose nanowhiskers. *European Polymer Journal,* Vol.48, No.3, pp. 454-463.

[57] Spagnol, C., Rodrigues, F. H. A., Pereira, A. G. B., Fajardo, A. R., Rubira, A. F., & Muniz, E. C. (2012b). Superabsorbent hydrogel composite made of cellulose nanofibrils and chitosan-graft-poly(acrylic acid). *Carbohydrate Polymers,* Vol.87, No.3, pp. 2038-2045.

[58] Sturcova, A., Davies, G. R., & Eichhorn, S. J. (2005). Elastic modulus and stress-transfer properties of tunicate cellulose whiskers. *Biomacromolecules,* Vol.6, No.2, pp. 1055-1061.

[59] Tashiro, K., & Kobayashi, M. (1991). Theoretical Evaluation of Three-Dimensional Elastic-Constants of Native and Regenerated Celluloses: Role of Hydrogen-Bonds. *Polymer,* Vol.32, No.8, pp. 1516-1530.

[60] Vandamme, E. J., De Baets, S., Vanbaelen, A., Joris, K., & De Wulf, P. (1998). Improved production of bacterial cellulose and its application potential. *Polymer Degradation and Stability,* Vol.59, No.1-3, pp. 93-99.

[61] Wang, B., Cai, Q., Zhang, S., Yang, X. P., & Deng, X. L. (2011). The effect of poly (L-lactic acid) nanofiber orientation on osteogenic responses of human osteoblast-like MG63 cells. *Journal of the Mechanical Behavior of Biomedical Materials,* Vol.4, No.4, pp. 600-609.

[62] Wu, Y. T., Zhou, Z., Fan, Q. Q., Chen, L., & Zhu, M. F. (2009). Facile in-situ fabrication of novel organic nanoparticle hydrogels with excellent mechanical properties. *Journal of Materials Chemistry,* Vol.19, No.39, pp. 7340-7346.

[63] Xiang, C. H., Joo, Y. L., & Frey, M. W. (2009). Nanocomposite Fibers Electrospun from Poly(lactic acid)/Cellulose Nanocrystals. *Journal of Biobased Materials and Bioenergy,* Vol.3, No.2, pp. 147-155.

[64] Zhang, Y., Yu, J. R., Zhou, C. J., Chen, L., & Hu, Z. M. (2010). Preparation, Morphology, and Adhesive and Mechanical Properties of Ultrahigh-Molecular-Weight Polyethylene/SiO2 Nanocomposite Fibers. *Polymer Composites,* Vol.31, No.4, pp. 684-690.

[65] Zhou, C. J., Qiu, X. Y., Zhuang, Q. X., Han, Z. W., & Wu, Q. L (2012a). In situ polymerization and photophysical properties of poly(p-phenylene benzobisoxazole) /multiwalled carbon nanotubes composites. *Journal of Applied Polymer Science,* Vol.124, No.6, pp. 4740-4746.

[66] Zhou, C. J., Chu, R., Wu, R., & Wu, Q. L. (2011a). Electrospun Polyethylene Oxide/Cellulose Nanocrystal Composite Nanofibrous Mats with Homogeneous and Heterogeneous Microstructures. *Biomacromolecules,* Vol.12, No.7, pp. 2617-2625.

[67] Zhou, C. J., Wang, Q. W., & Wu, Q. L. (2012b). UV-initiated crosslinking of electrospun poly(ethylene oxide) nanofibers with pentaerythritol triacrylate: Effect of irradiation time and incorporated cellulose nanocrystals. *Carbohydrate Polymers,* Vol.87, No.2, pp. 1779-1786.

[68] Zhou, C. J., Wang, S. F., Zhang, Y., Zhuang, Q. X., & Han, Z. W. (2008a). In situ preparation and continuous fiber spinning of poly(p-phenylene benzobisoxazole) composites with oligo-hydroxyamide-functionalized multi-walled carbon nanotubes. *Polymer,* Vol.49, No.10, pp. 2520-2530.

[69] Zhou, C. J., Wang, S. F., Zhuang, Q. X., & Han, Z. W. (2008b). Enhanced conductivity in polybenzoxazoles doped with carboxylated multi-walled carbon nanotubes. *Carbon,* Vol.46, No.9, pp. 1232-1240.

[70] Zhou, C. J., & Wu, Q. L. (2011). A novel polyacrylamide nanocomposite hydrogel reinforced with natural chitosan nanofibers. *Colloids and Surfaces B-Biointerfaces,* Vol.84, No.1, pp. 155-162.

[71] Zhou, C. J., Wu, Q. L., Yue, Y. Y., & Zhang, Q. G. (2011b). Application of rod-shaped cellulose nanocrystals in polyacrylamide hydrogels. *Journal of Colloid and Interface Science,* Vol.353, No.1, pp. 116-123.

[72] Zhou, C. J., Wu, Q. L., & Zhang, Q. G. (2011c). Dynamic rheology studies of in situ polymerization process of polyacrylamide-cellulose nanocrystal composite hydrogels. *Colloid and Polymer Science,* Vol.289, No.3, pp. 247-255.

[73] Zhou, C. J., Yang, W. M., Yu, Z. N., Zhou, W. L., Xia, Y. M., Han, Z. W., & Wu, Q. L. (2011d). Synthesis and solution properties of novel comb-shaped acrylamide copolymers. *Polymer Bulletin,* Vol.66, No.3, pp. 407-417.

[74] Zhou, C. J., Zhuang, Q. X., Qian, J., Li, X. X., & Han, Z. W. (2008c). A simple modification method of multiwalled carbon nanotube with polyhydroxyamide. *Chemistry Letters,* Vol.37, No.3, pp. 254-255.

[75] Zoppe, J. O., Peresin, M. S., Habibi, Y., Venditti, R. A., & Rojas, O. J. (2009). Reinforcing Poly(epsilon-caprolactone) Nanofibers with Cellulose Nanocrystals. *ACS Applied Materials & Interfaces,* Vol.1, No.9, pp. 1996-2004.

Chapter 7

# Semiconductor Nanocrystals

Anurag Srivastava and Neha Tyagi

Additional information is available at the end of the chapter

http://dx.doi.org/10.5772/50440

## 1. Introduction

In terms of continuous miniaturization of electronic devices nanotechnology is a ray of hope for semiconductor industries due to peculiar properties of nano-materials that changes significantly the efficiency of the devices. Research in nanoscale materials get started because of the unique properties that are obtain at this scale, by changing the shape or size of these materials. At nanoscale, the behavior of materials drastically get changed and hence their properties. Particularly in semiconductors, it results due to the motion of electron to a length scale that is equal or smaller to the length scale of the electron Bohr radius that is generally a few nanometers. Continuous efforts are being made to explore the new physical properties of materials and to engineer them to fit for various technological applications. Scientific community has paid much attention to study various aspects of variety of nanostructured materials like; fullerenes, nanotubes (NT), nanoribbons (NR), nanowires (NW) and most recently, the nanocrystals (NCs). As the name indicates, NCs are the structures exhibiting crystalline structure but with one, two or all the three dimensions within the range of 1-100 nm. NCs enjoys advancement of crystalline periodicity at nano regime ($10^{-9}$ m) and often possesses new properties, most of the time reverse from those of the equivalent bulk materials due to their large surface to volume ratio, quantum size effect, and confinement effects [1-3]. Quantum confinement effect is the main phenomena that often been observed in NC, which deals with the spatial inclusion of the electronic charge carriers within the NC. The quantum size effect is analogous to the well known quantum mechanical problem i.e when a particle is place in a box, where the energy partition between the levels raises as the dimensions of the box reduces. Due to the same, we can observe an increment in the energy band gap of the semiconductor as the size of the crystal reduces. Nanostructuration is not confined to human imagination only; in fact it's an important choice of nature. The word nanotechnology is not new for the nature, where it has been realized for over billions of years which can be for instance in clays, circumstellar, dusts or many biological systems too. One of the most interesting examples of nature inspired nanotechnology is the water proof leaves of lotus owing to which a lotus can sustain in water for a very long time. Recently, it has been verified that lotus leaves has a special coating of nanoparticles which behaves as water repellant, this water repellant property has motivated number of researchers to perform research in this area and some of them even succeeded to manufacture hydrophobic coatings [4]. Interaction at the nanoscale are usually governed by the fact that the characteristic nano-length becomes comparable with other critical lengths of the system as mean free paths, scattering or coherence lengths. At the same time, confinement translates into a reorganization of the electronic density of states towards more distinct states. These two consequences of nanostructuration and their combination lead to the so called quantum related effects which determines most of the peculiar properties of nano-materials. A major corollary of nano confinement is the elevated ratio of the number of surface to volume atoms, which becomes another key point for the understanding of these extraordinary properties. In small NCs the ratio of the number of surface to the total number of atoms is less than one, for which it scales as ~1/(Particle diameter)$^3$ and exactly one in nanotubes or fullerenes, whereas in bulk crystal it goes to $10^{-20}$ or less.

NCs are very tiny crystals that results when a single crystal experience a solid to solid phase transition. Generally, NCs can be prepared by a huge number of atoms presented on the surface and no interior defects, unlike bulk materials. We all know that under the application of pressure abrupt change in the arrangement of the atoms i.e. structural transformation from one phase to another has been observed in crystals. The understanding of pressure-induced phase transition in various nanomaterials plays an important role in probing the properties of new materials. The consequences of finite crystal size on the structural phase transition and bulk modulus as well as compressibility has widely been demonstrated to get better understanding towards the stability, electronic, mechanical as well as other properties of nano-materials. Difference in surface energies of the two crystal phases is mainly responsible for the change in transition pressure in case of nano-materials. In fact the general rule states that smaller the size of the crystal, higher the transition pressure has been verified by several systems [5,6]. A report shows that NC shape can be easily observed using transmission electron microscopy at atmospheric pressure [7].

## 2. Applications

The enormous potential of exploring the new sciences and technology at nanoscale, may impact on industrial productivity, realized all over the world to develop the new materials for variety of newer applications. NCs possess variety of applications in the field of electronics, opto-electronics and photonics, and also several biological (medical) treatments following are the few important applications of semiconductor NCs.

### 2.1. Nanocrystals LEDs

In the last two decades very basic and more efficient white light-emitting diodes (LEDs) have been formed by replacing phosphors (a colour converter) with CdSe based NCs that are integrated directly into the p-n junction [8]. CdSe NCs/nanorods emit linearly polarized light along the crystallites axis and the degree of polarization depends upon the surface to volume ratio of the NCs [9]. A hydride structure comprises of single layer of CdSe NCs accumulate at the top of an InGaN/GaN quantum well (QW) has been utilized as LED which provides ~10% color conversion efficiency. An inverted LED design is shown in Fig. 1(a),where an InGaN QW grow on the top of a thick p-doped GaN barrier and complete the structure with a thin n-type GaN cap layer (normally a QW is grown on top of the n-type layer). Electrical contacts have been deposited by electron beam evaporation using an inter-digitated mask that comprised of several different device-mesa and contact geometries, as shown in Fig.1(b) [10].

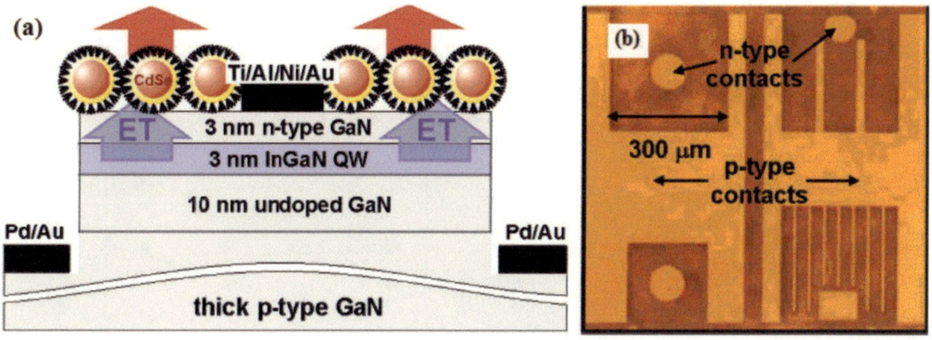


**Figure 1.** (a) Schematic of the nonradiative energy transfer (ET) LED structure. (b) Contact geometry of real device Ref.[10]

The conjugated polymers and InAs based NCs have been utilized to produce near infrared LEDs, where the emission can be tuned from the range of 1 to 1.3μm that efficiently covers the short-wavelength telecommunications [11]. Semiconductor NCs when excited electrically using GaN injection layer can be used as color selectable chromophores, shows high emission efficiencies with excellent photo stability and chemical flexibility. In a report by Mueller et al. [12] multicolour LEDs have been demonstrated where semiconductors NCs are integrated into a p-n junction made by GaN injection layers. Si NCs LEDs have been fabricated using ion implantation and observed the enhancement in LEDs optical output power (~20 nW), with typical power efficiencies ranging from 1 to 10% [13]. Light emitting transistors integrated with individual CdSe NCs shows occurrence of Coulomb blockade at low bias voltage and low temperature, which signifies that electrons pass through the NC by single electron tunneling. Fig.2(a), shows a single nanorod contacted by two Au electrodes with about ~30 nm separation. The representative current-voltage and electroluminescence (EL) data collected from a device "D1" is illustrated in Fig.2(b). A nonlinear increase in current ($I$) at high bias voltage ($V$) clearly reveals that the device exhibits a low-bias conductance gap. Once $V$ reaches a threshold ($V_{th}$) the device starts emitting light as shown by EL intensity measurement [14].

## 2.2. Nanocrystal memories

Floating gate structure is firstly invented by Sze and Kahnd in 1967, which was used to construct flash memories [15] (Fig.3a). In 1995, IBM proposed first discrete NC memory and in early 2000's scientists have considered NCs as the capable candidate that can solve the current scaling problem (Fig.3b). The metal-oxide-semiconductor (MOS) memory structures based on Si NCs have potential applications in flash memory, reason being Si NCs are implanted as charge storage nodes in an oxide layer between the control gate and tunneling layer that reduces the difficulty of charge loss which can generally seen in conventional flash memories. Si NC provides faster write/remove speeds, small injection oxides, little operating voltages and better stamina [16]. Literature confirms that the information is stored in the NCs by injecting charges, due to which a transistor needs much voltage for turning it ON, known as program operation illustrated in Fig.3(c). If we apply read voltage ($V_{read}$) to the gate between the program and erase operations to read the corresponding drain current ($I_d$) then we obtain the memory status given by values 0 and 1 [17]. Generally the quantum dot/NCs flash memories have used Si NCs as a substitute to floating gate layer, however many results demonstrates the superiority of Ge based NC memories over those based on Si, because Ge NCs provides large nonvolatile charge conservation time because of their small band gap.

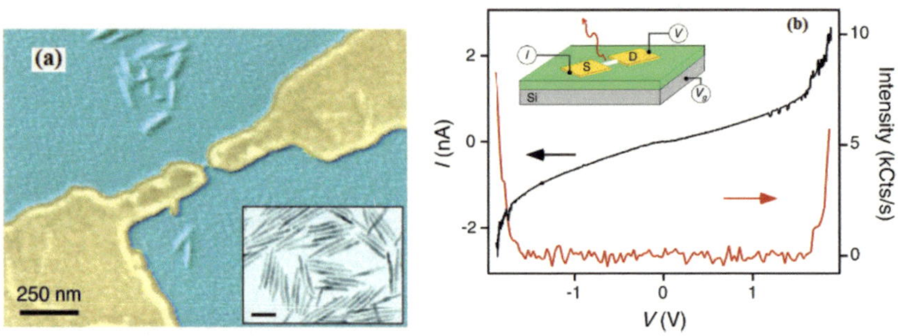


**Figure 2.** (a) SEM image (false color) of CdSe nanorod transistor. The inset shows a TEM image of the CdSe nanorods (scale bar is 50nm) (b) Current (black) and concurrently calculated EL intensity (red) plotted against bias voltage obtained from device. Inset illustrates the transistor model, with source (S) and drain (D) electrodes connecting the nanorod with a back gate is shown [Ref.14].

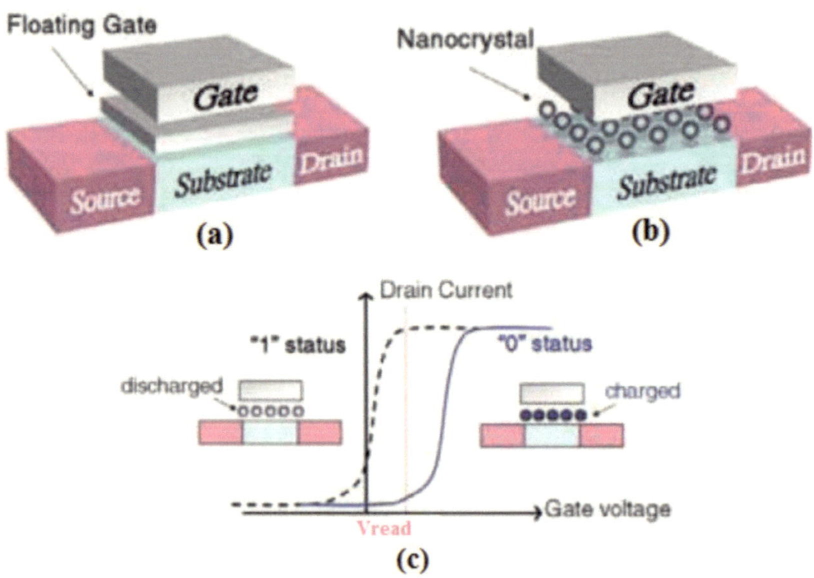


**Figure 3.** (a) Floating gate non volatile memory structure. (b) NC non volatile memory structure. (c) Program and erase mode of the NC memory device Ref.[17].

NCs may work as memory cells, where the conventional poly-silicon floating gate is replaced by an array of Si NCs, as the single cell and cell array of 1 Mb and 10 k have been realized by using a conventional 0.15 μm FLASH technology [18,19]. Ge NCs have been fabricated by pulsed laser deposition for their use in floating gate memory, such memories shows excellent charge preservation characteristic. [20].

## 2.3. Nanocrystal photocharge generators

The PbSe NCs represent a fascinating system due to the easiness of knowing the quantum modulated optical nature in the infrared range with a Bohr exciton radius of PbSe around 46 nm, where the quantum confinement effect come into view at comparatively large particle size. As we know that PbSe bulk crystal has rocksalt type phase and a direct band gap semiconductor ($E_g$ = 0.28 eV) whereas the PbSe NC show well defined band-edge excitonic transition, tuned from 0.9 to 2.0 eV [21], leads to an efficient photocharge generators for communicating the IR wavelengths [22]. Highly efficient photo-detectors based on composites of the semiconducting polymer and PbSe NCs have been prepared and the outer quantum efficiency in these devices is larger than one [23]. Ge NCs photo-detector possesses powerful optical absorption and its photocurrent response has been measured within the wavelength range of 1.3 to1.55 μm with a low dark current of 61.4 nA along with a a photocurrent responsivity of 56 mA/W at the 5 V reverse bias [24].

## 2.4. Nanocrystal solar cells

An InAs/GaAs quantum dot solar cell has been synthesized and observed superior photocurrent without reduction of open circuit voltage compared to a solar cell without quantum dots. These solar cells have a light absorption range, which is extended up to 1.3 μm and confirms a trade-off between open circuit voltage and quantum dot ground state energy [25]. An ultrathin solar cell composed of CdSe and CdTe NCs of size 40nm shows 3% power conversion efficiency and the device remain stable in air [26] (Fig.4).

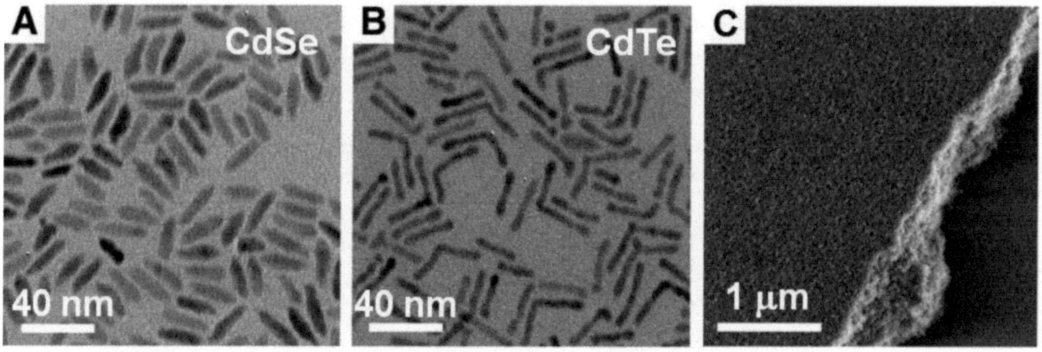


**Figure 4.** Transmission electron micrographs of (A) CdSe and (B) CdTe NCs (C) Spin-cast film of colloidal NCs imaged by scanning electron micrography is uniform and without defect; the film edge of this ~100 nm film is shown for difference with the Si substrate Ref[26].

An induce electric field is large enough when two different molecules attached to a single NC and the field significantly changes the electronic as well as optoelectronic properties of the NC. By the help of mixed ligand, induced electric field significantly increases the efficiency of charge generation in CdSe NC solar cells, which enhanced the complete cell efficiency [27]. Si NC have quantum confinement property and so it can be used as photovoltaic materials since its ability to collect the photo-generated current through efficient electronic transport and for the development of Si NC based solar cells the exciton dissociation is a recent challenge [28].

### 2.5. Nanocrystals in health

Enzymes, membranes, nucleic acids, etc. are the elementary functional units made up of complex nanoscale particles in biological systems. In the era days of miniaturization where metals, semiconductors and magnets through which we construct optical and electrical sensors, can now be prepared on the scale of individual biological macromolecules that will have large impact on forthcoming medical treatments. In drug delivery and clinical applications, nanotechnology is one of the key factors for modern drug therapy. Due to the simplicity in preparation and general utility NCs are the new carrier-free colloidal drugs delivery systems having a particle size of 100-1000 nm, these NCs have been thought of as a practical drug delivery approach to develop the poorly soluble drugs. Literature shows that the number of drugs coming directly from synthesis is nowadays poorly soluble and the drugs which are very less soluble in water combine very poorly with bioavailability. If we do not get any way to improve drug solubility then it will be very difficult to get absorbed from the gastrointestinal zone in the bloodstream and then to reach at the proper site for required action [29,30]. Recently intense research is being made on the colloidal quantum dots due to their stable light emitting nature, which can be broadly tuned by varying the size of the NC. Lots of efforts have been devoted in last two years on the development of large range of methods for bio conjugating colloidal NCs [31] because of their diverse applications such as in vivo imaging [32], cell tracking [33], DNA detection [34] etc. While studying the optical properties of NCs, it has been noticed that NCs fluorescence wavelength sturdily depends on their size as a result of which NCs photo bleach property get reduced. These NCs are fascinating fluorescence probes for various types of labelling research like cellular structures labeling, tracking the path and absorption of NCs by living cells [35].

## 3. Synthesis & characterization

In general the performance of the material depend on its properties whereas the properties depend on the crystal geometry, constituents, defects and interfaces that are definitely prohibited by thermodynamics and kinetics of the fabrication process. In 1980's when the first theoretical explanation was proposed for colloidal spherical NCs by Brus [36], together with advances in the synthetic procedures [37] lead to a rapid increase in research in the field of nano sized materials. Following are the popular methods developed for the fabrication and characterizations of NCs:

**Solution-phase** methods are being used for preparing CdS NCs with a wide variety of morphologies, starting with oleylamine as capping agent and by varying the reaction conditions [38]. Low temperature inverse micelle **solvo-thermal route** method has been used to synthesis large amount of single crystalline Ge NCs and X-ray diffraction measurement illustrates that these NCs are composed of pure cubic Ge structure. The size and morphology of the NCs are observed by transmission electron microscopy (TEM). In Fig.5(a) a TEM image of Ge NCs is illustrated and Fig.5(b) shows selected area electron area diffraction (SAED) sample for the NC of 25 nm size and displays the typical spot pattern of a single crystal domain and also the lattice planes of Ge with [111] hexagonal symmetry. High resolution TEM (HRTEM) images of high quality Ge NCs are shown in Fig. 5(c) and (d) [39].

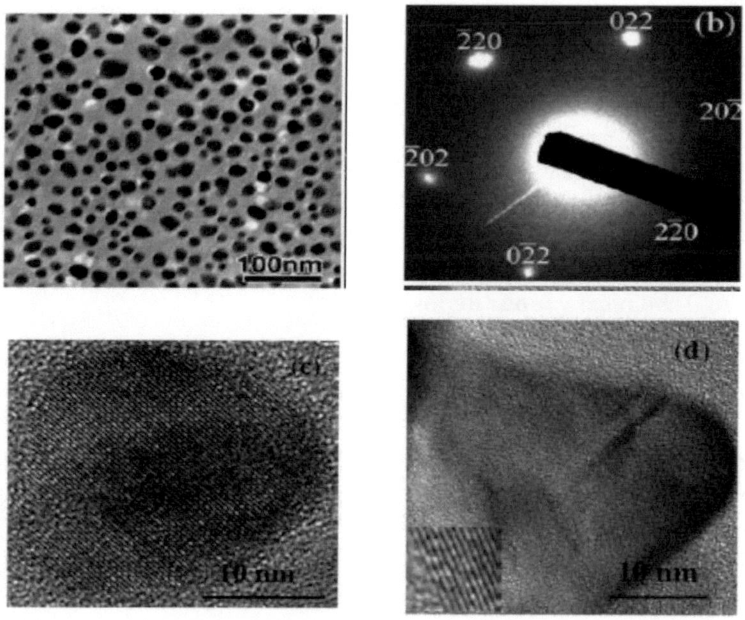


**Figure 5.** (a) As-prepared Ge NCs TEM image, (b) SAED sample of Ge NC with 25nm size; HRTEM images of spherical (c) and triangular(d) Ge NCs Ref.[39]

Few **non-hydrolytic methods** have also been developed for the synthesis of ZnO NCs with controlled size, shape and surface-defect at low temperature. But such methods need very cautious control of conditions and carefully engineered precursor.

**Direct liquid phase precipitation** is a new process proposed recently for NC fabrications, however slight hard from the previous complex fabrication methods in superficial and efficient way (at room temperature). This process produced perfect ZnO NCs of different diameter (5 to 12 nm) and of various shapes [40].

**Benzene thermal** method at low temperature and low pressure (0.12–0.14 MPa) has been used for the synthesis of AlN NCs. Owing to the catalytic effect of AlN nanocrystals, ~8% benzene was converted at 150°C into several polymers, like cyclohexylbenzene [41].

**Sol-gel** technique can be used for the fabrication of amorphous or crystalline materials from a liquid phase at low temperature. This technique employed for the production of laser dyes, enzymes, nano sized semiconductors and metal nanoparticles. AlP NCs embedded in silica glasses have been prepared by sol gel process, where the preparation of gels take place by the complex solution hydrolysis and after about 10hours heating in an air atmosphere to form gel glass, again get heated in the presence of $H_2N_2$ gas and react with Al(III), finally a cubic AlP crystallites are produced [42]. Similar technique has been used in producing InAs and InP microcrystallites embedded in $SiO_2$ gel glasses [43,44]. GaAs NCs have been synthesized by an **electrochemical route** method from the acidic solutions of metallic gallium and arsenic oxide [45]. The growth of 1D nanostructures have traditionally been associated with such highly technological processes/analysis tools as **vapor deposition**, **lithography**, **thermally assisted electromigration** method [46], **annealing process** [47], scanning tunneling microscope (STM), atomic force microscope (AFM) [48]. By using size selective **precipitation techniques**, cadmium chalcogenide NCs have been successfully obtained [49]. The most common semiconductor materials grown by precipitation technique are CdTe, CdSe, CdS, and ZnS of which CdTe has special technological importance because it is the only known II-VI material that can form conventional p-n junctions [50]. For the synthesis of III-V compound semiconductors on the semiconductor substrate variety of methods have been developed starting from **metaorganic chemical vapour phase epitaxy** (MOVPE), **molecular beam epitaxy** (MBE) [51], various **inorganic techniques** [52], **porous glass** [53] to photo **chemical vapour deposition** [54].

In **chemical vapour deposition (**CVD), the vaporized precursors are introduced into a CVD reactor and adsorbed onto a substance held at high temperature. These adsorbed molecules will either thermally crumble or react with other gases to form crystals. The CVD process has been organised in three steps (i) mass transport of reactants to the growth surface through a boundary layer by diffusion (ii) chemical reactions on the growth surface, and (iii) removal of the gas-phase reaction by products from the growth surface. NCs have been characterized by X-ray diffraction, scanning transmission electron microscopy (STEM), optical absorption spectroscopy and fluorescence spectroscopy.

Keeping in view of all the techniques discussed for the synthesis and characterization of NCs we can state that in current scenario it has been easier task to fabricate a NC of desired morphology.

## 4. Properties

### 4.1. Structural properties

The structural properties of nanocrystals have been performed using various experimental as well as theoretical approaches. In general, a structural property means the analysis of stability of the crystal structure and then the evaluation of ground state parameters such as bond length, bond angle, lattice parameter etc. For example a periodic cluster approach with an atomistic pair potential has been employed to simulate AlN NC and predictes that the NCs displays graphitic-like layers and non-buckled wurtzite structure. The variation of surface tension, *c/a* ratio, and equation of state for AlN NC with respect to both the pressure and size has also been examined. The study also reveals that the bulk modulus of the wurtzite phase of NC is smaller (although decreasing with size) than its bulk counterpart [55]. Recently, Franceschetti [56] has analyzed the structural and electronic properties of PbSe NCs with different radius. Fig.6 shows the optimized structures of PbSe NCs in $Pb_{44}Se_{44}$, $Pb_{140}Se_{140}$ and $Pb_{240}Se_{240}$ form and both Pb and Se atoms have rocksalt structure. The Pb-Se bond lengths and bond angles are appreciably distorted contrast to that of its bulk counterpart as represented in Fig.7. In fact, Fig.7 clearly shows the deviation of bond length of the Pb-Se with respect to the distance of the bonds from the center of the NC and the bond lengths of Pb-Se NCs are in close match with its bulk crystal bond length (3.103 Å). The average Pb-Se bond length of the NC in the ground state is smaller than the bulk crystal bond length and decreases with the size. The formula for calculating formation energy of NC (PbSe) can be given as

$$\Delta E_F = E_{NC} - NE_{bulk} \tag{1}$$

where the calculated total energy for NC is $E_{NC}$ with calculated bulk crystal total energy $E_{bulk}$ and total number of Pb and Se atoms are represented by N for NC.

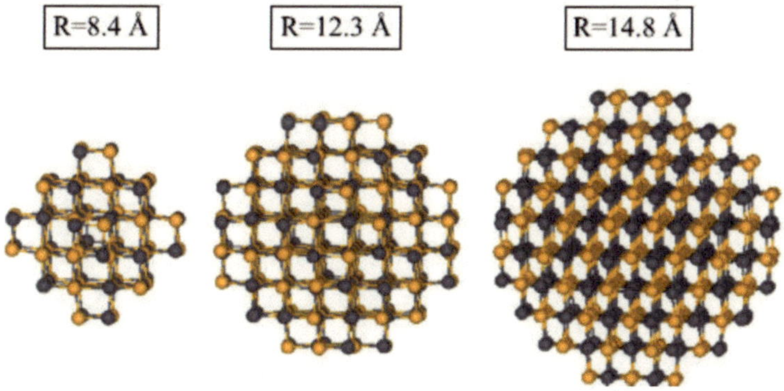


**Figure 6.** Three different sizes optimized geometries of PbSe NCs. The dark circles indicate Pb atoms and the light circles for Se atoms [56].

Srivastava et al. have also investigated the structural properties of various bulk as well as NCs such as AlAs [57,58], AlSb [59] and AlN [60]. The structural properties of bulk AlAs in zincblende (B3), wurtzite (B4), NiAs (B8), CsCl (B2) and NaCl (B1) type phases have been analyzed through first principle density functional theory (DFT) approach. Using local density approximation ( LDA) with Perdew-Zunger (PZ) type parameterization the equilibrium lattice constant *(a)* of original B3 type phase of bulk AlAs is calculated as 5.64 Å and also from GGA-PBE (5.72 Å) and GGArevPBE (5.77 Å). The stability analysis of bulk AlAs in B8 type phase, corresponds to the *c/a* ratio of 1.59 Å, where the equilibrium lattice constants *a* and *c* are 3.72 Å and 5.91 Å, respectively [57]. Similar approach has been used for analyzing the structural stability of AlAs NCs in B3, B4, B2 and B1 type phases at ~0.9 nm, where the lattice parameters for all the four stable phases of AlAs NCs [58] such as B3, B4, B1 and B2 are 5.761, 4.283, 5.350 and 3.207 Å, respectively. The total energy values for the B3, B4, B1 and B2 type phases of AlAs NCs are −468.52, −468.46, −467.84 and −467.77 eV, respectively. Fig.8 shows the variation of energy with respect to volume for B3 and B1 type phase of AlAs NCs. The calculated lowest total energy confirms that the B3 phase as the most stable one at lower pressure and the B1 type phase can be considered as the high pressure phase. On comparison with bulk AlAs, stability trend of the low pressure phases are same in bulk and nano-dimension but the high pressure phases are different.

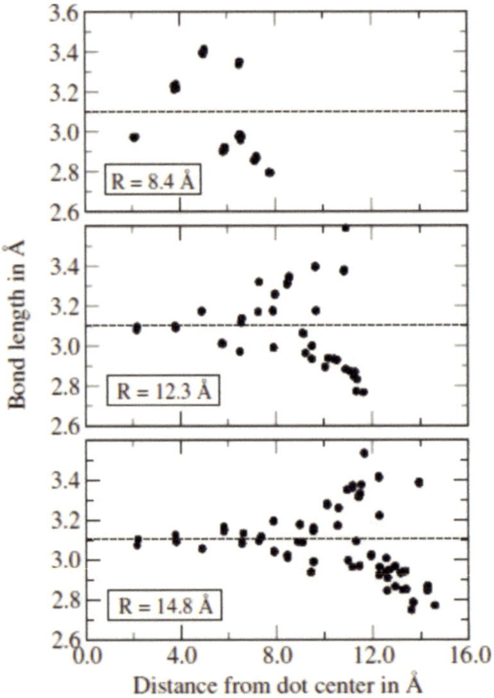


**Figure 7.** The variation of all Pb-Se bonds as a function of distance of the bond from the geometric centre of the NC. The dash lines shows the computed bulk Pb-Se bond length Ref.[56].

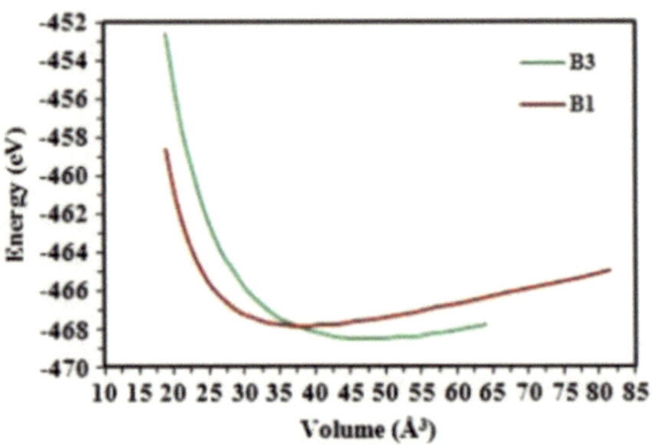


**Figure 8.** Energy as a function of volume for AlAs nanocrystals Ref.[58].

The stability of AlSb nanocrystal of ~1 nm size has also been analyzed in B3, B1 and B2 type phases theoretically. The article reports that computed lattice constant (6.235 Å) for bulk

AlSb in B3 type phase is in close match with its experimental as well as other theoretical counterparts. However the computed lattice constants of B3, B1 and B2 type phase of AlSb NCs are 5.33 Å, 5.92 Å and 3.57 Å respectively. The findings conclude that B3 type phase is most stable with lowest total energy and highest binding energy [59]. In an another report by Tyagi et al. ab-initio pseudopotential approach has been used for calculating the lattice constants of 1-D AlN NC in B3, B1 and B2 type phase as 4.58 Å, 4.48 Å and 2.98 Å respectively [60]. Besides these theoretical works few experimental work has also been performed as zinc-blende GaP NCs have been synthesized by three different methods and it has been found that NCs transferred irreversibly into four new phases of GaP NCs with different lattice constants (5.8872, 5.5493 and 7.0339 Å) and the largest one is about 1.3 times of the reported data, which has been characterized by X-ray diffraction and transmission electron microscopy [61].

## 4.2. Electronic properties

Electronic structures in nano regimes are extremely enviable for the future potential electronic devices. On the other hand, nano structures are still a challenge, where different experimental as well as theoretical approaches have been employed to understand the variation of electronic structures of NCs as a function of its size. The study of quantum confinement phenomena in semiconductor NCs is the subject of intense research which in fact determines their electronic behavior. A theoretical approach effective mass approximation (EMA) [62] is employed firstly to understand quantum confinement effects on the electronic band gap as a function of size of NCs. The confining potential for the electron and hole are assumed infinite in most of the EMA calculations. In EMA approach, the wave function of both electron as well as hole get vanishes at and beyond the surface of the NCs without any tunneling. To date numerous theoretical methods have been developed such as first-principles [63], semiempirical pseudopotential [64] and tight binding (TB) method [65] to see the effect of size on the electronic band structure of various NCs. The TB method has received great attention because of its realistic description of structural and dielectric properties in terms of chemical bonds and ease of handling large systems. For the description of hole wave function the effective bond orbital model and EMA calculations are used for of electron wave function by Einevoll [66], and Nair and his coworkers [67]. To see the effect of NC size on the electronic band structures various pseudopotential and TB calculations have been performed. UV-visible absorption spectroscopy tool has been utilized to study the variation of band gap as a function of NC size. There are reports on the individual shifting of highest occupied molecular orbital of the valence band and the lowest unoccupied molecular orbital of the conduction band region with NC size, using various methods of energy spectroscopy e.g. X-ray absorption and photoemission [68]. Another semi empirical pseudopotential method has been employed to calculate the electronic structures of Si, CdSe [69] and InP [70] nanocrystals. An appreciable progress has been observed in the accuracy and the results of the electronic band structure calculations of NCs while using TB and pseudopotential approaches. In comparison to EMA, the reason behind the popularity of TB might be low computational effort, basic picture of atomic orbitals and hopping interactions for a predefined range. The ab-initio electronic band structure computation may lead to perceptive TB scheme and show accuracy because it use physical and real basis set. In 1989 Lippens and Lannoo [71] have used semi-empirical TB approach with $sp^3s^*$ orbital basis to compute the variation in the energy gap of semiconductor NCs as a function of size and the results so obtained are in good agreement with the EMA potential, however the $sp^3s^*$ TB model have a tendency to underestimate the energy gap. This model cannot reproduced the lowest lying conduction band, however the electron-hole interaction improves the nearest neighbor $sp^3s^*$ model by the inclusion of spin-orbit coupling. In a study on InP [72] NCs where the $sp^3d^5$ orbital basis is used for the anion and the $sp^3$ basis for the cation with the next nearest neighbor interactions, gives good agreement with the experimental values. In $sp^3d^5$ orbital basis addition of the next nearest neighbor interactions for the anion gives correct explanation of the electronic band structure of semiconductors in its bulk as well as NCs [73]. Diaz et al.[74] have studied the electronic structure and optical spectra of GaAs NCs for a wide range of sizes by using both $sp^3s^*$ and $sp^3s^*d^5$ nearest-neighbor TB models and found that the enclosure of *d* orbitals into a minimal basis set is essential for a good explanation of the lowest lying states of electron, particularly in the well confined system. Franceschetti [56] has analyzed the electronic properties of PbSe NCs within the radius range of 8.4-14.8 Å. Wannier function approach has been employed to study the band gap dependence on the different shapes (spherical, cubical, and tetrahedral) of CdS NCs and the report suggests that energy gap depends on the number of atoms present in the NC and not on their arrangements. Fig.9 shows the variation of energy gap variation as a function of the number of unit cells in the NC of different shapes, indicates that the energy gap slightly depends on the arrangements of the atoms in the NC [75].

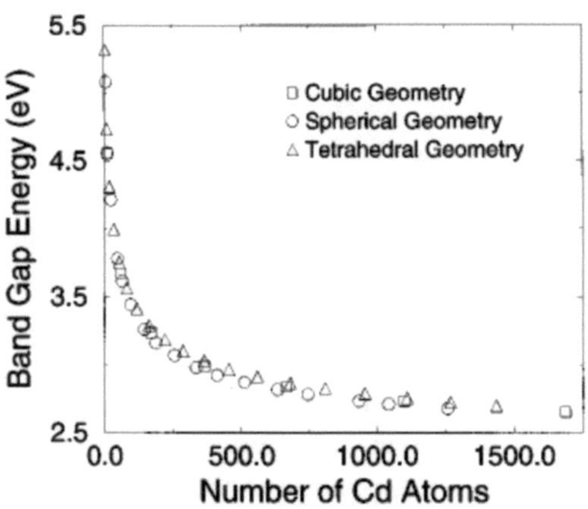


**Figure 9.** Dependence of band gaps on the number of Cd atoms for various shapes NC. Ref.[75].

Using first-principle DFT approach Srivastava et al. have analyzed the electronic properties of various bulk/NCs such as AlN, AlSb, AlAs [57-60] etc. in their different structural phases. The detailed discussion of the electronic structures of 1-D AlN NC in its various possible phases like zinc blende (B3), NaCl (B1) and CsCl (B2) is as follows. The band structure of B3 type phase of AlN NC as shown in Fig.10(a) has a band gap of about 1.6eV at Γ point, defends the semiconducting nature. Fig.10(b) shows the band structure plot for B1 type phase of AlN NC, clearly depicts a band gap of around 1.5eV at Γ point however the B2 type phase is purely metallic in nature. The band gaps of all the AlN 1-D NCs are quite smaller as compared to bulk crystal, which contradicts to the quantum size effect where usually enhancement in the band gaps has been observed. This abnormal AlN band gaps is due to the surface effects. The surface states related with the tri coordinated N and Al atoms at lateral facets in NC creates new energy levels at valence band edges as well as conduction band edges that directly narrows the band gaps relative to their bulk counterparts.

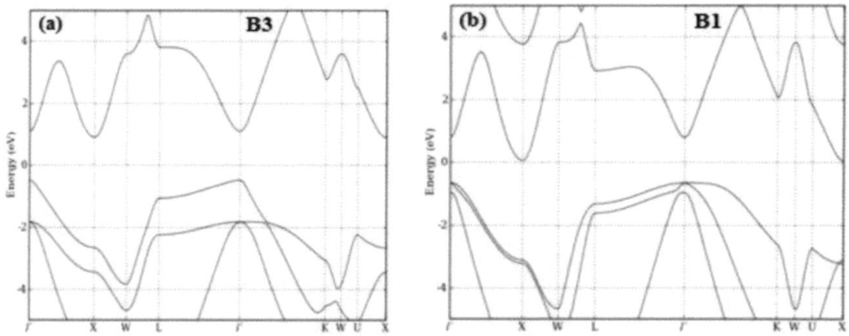


**Figure 10.** (a-b) Electronic Band structure of AlN NC in B3 and B1 type phase Ref.[60].

The density of states for the most stable B3 type phase of AlN NC has also been analyzed in the energy range of -5.0 to 5.0 eV shown in Fig. 11, with three visible distinct peaks, where a highest peak appears in the valence band region near -2.3eV and satisfies the band structure plot by showing the unavailability of peaks at the Fermi level.

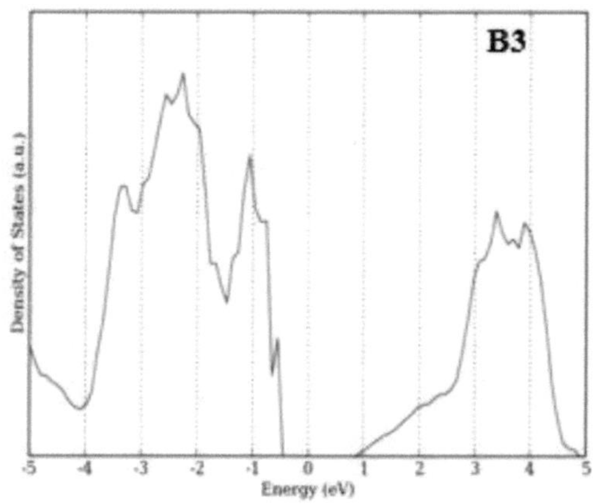


**Figure 11.** Density of states of AlN NC in different structural B3 phase Ref.[60].

### 4.3. Optical properties

It is well known that the band gaps of different semiconductors such as III-nitrides, II-VI group, IV group elements can be tuned by varying the composition in their alloys which make these systems prominent for various the applications like in optoelectronic devices, to use them for a spectral range starting from deep infrared to distant ultraviolet region. Quantum confinement has a great impact on the optical properties of semiconductor NCs. The phenomena of absorption and emission in the semiconductors are governed by their fundamental direct band gap. The increment in the energy band gap and strengthening of the oscillations with decrease in size of NCs has been revealed by spectroscopic studies. Due to the advancements in the synthesis, characterization and size selections of direct band structures of III-V and II-VI semiconductors, the spectroscopic studies of quantum confinement effects in these systems become much easier. The potential candidature of semiconductor NCs as light absorbers in solar-cell devices of third generation, has attracted number of researchers. For example PbSe, PbS, CdSe, PbTe, InAs and Si NCs shows much efficient multiple-exciton generation i.e. the chance of producing multiple electron-hole pairs from a single, high-energy photon, which thus increased the solar cell power [76]. For the solar energy conversion the semiconductor NCs requires a good match between the NC absorption spectrum and the solar spectrum. Keeping in mind that the quantum confinement effect tends to open up the energy band gap, the narrow band gap semiconductor NCs such as Ge, InAs, group-III antimonides and lead chalcogenides [77] are the potential candidates for the solar cell applications. Due to indirect band gap bulk Si is a poor light emitting material. In Si the excited electron-hole pairs have very wide radiative lifetime (ms), hence opposing non-radiative recombination rule and cause most of the excited electron-hole pairs to recombine non-radiatively. Si NCs formed by aerosol technique or by thermal precipitation of Si atoms rooted in $SiO_2$ layers, give strong visible and near infrared photoluminescence (PL), similar characteristics as in porous Si [78]. Optical properties of isolated Si NCs of ~4 nm size have been analysed with spectroscopic ellipsometry in the photon energy range of 1.1-5.0 eV. Si NCs reveals a major reduction in the dielectric functions and optical constants and whereas shows a large blue shift (~0.6 eV) in the absorption spectrum. The dielectric function and optical constants of the Si NC are shown in Fig. 12 and 13, where overall spectral features of Si NC are similar to bulk Si. Nevertheless, the Si NC shows a remarkable drop in the optical constants and dielectric function in contrast to bulk Si. It is well recognized that decrease in the static dielectric constant becomes significant as the size of the structure approaches to the nano regime [79].

Photoluminescence (PL) properties of GaAs NCs in $SiO_2$ matrices formed by sequential ion implantation and thermal annealing, have been studied and observe broad PL due to GaAs NCs appears in the red spectral region, where the spectral shape of the red PL band depends on the hydrogen concentration in the sample. These GaAs NCs in $SiO_2$ film with low defects density will be used as the material for the future excitonic devices [80]. The blinking (or ON-OFF) behavior has been observed in single CdSe NC PL, which gets attributed to an intermittent photo ionization and subsequent neutralization of the NC [81]. The optical properties are related to the edge transition of the electronic band gaps in a semiconductor. Hence it is necessary to study the size dependence of the electronic band gap and the related exciton energy in semiconductor NCs. Few earlier experiments found that the lack of a large overlap between absorption and emission spectra in CdSe NC can improve the efficiency of light-emitting diodes (LEDs) due to the reduction in re-absorption

[82]. Photovoltaic response has been investigated in PbSe NCs of various sizes and conjugated polymers. The devices composed with these NCs shows good diode characteristics and sizable photovoltaic response in a spectral range from the ultraviolet to infrared [83]. A very interesting observation about the NCs is that their band gap can be tuned over a large energy range simply via synthetic control over the size of the NCs [84]. NC having absorption onset in the near to mid infrared (IR) will also strongly absorb solar photons of higher energy. Generation of multiexcitons in PbSe NCs from a single photon absorption event is observed to take place in picosecond and occurs with up to 100% efficiency depending upon the excess energy of the absorbed photon. Thermodynamic conversion efficiency in solar cells is found to be 43.9% in concentrated solar lighting. The computed percentage depends upon the hypothesis that the absorption of an individual photon with energy above a semiconductor band gap results in formation of a single exciton and the photon energy which is more than the energy band gap get lost by electron-phonon interactions. Numerous techniques have existed to raise the power exchange efficiency of solar cells together with the improvement of multi-junction cells, intermediate band devices, hot electron extraction and carrier multiplication [85,86]. In $Si_{35}$ and $Si_{66}$ cores passivated with oxide shows lowering of energy band gap by 2.4 eV and 1.5 eV. The oxide and hydrogen passivated NCs are optically forbidden indirect band gap and optically allowed direct band gap transitions respectively. However the highest occupied molecular orbitals and lowest unoccupied molecular orbitals are delocalized in both cases. Theoretical prediction get confirmed through the experimental observation that hydrogen passivated Si NCs emit blue light whereas oxide passivated Si NCs emit yellow-red light [87].

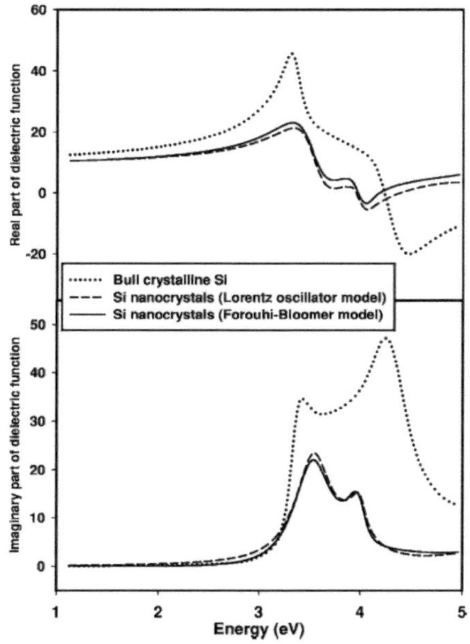


**Figure 12.** Real (ε1) and imaginary (ε2) parts of the complex dielectric function of the Si NC obtained from the spectral fittings based on the Lorentz oscillator model and the FB model. The dielectric function of bulk Si is also included for comparison Ref.[79].

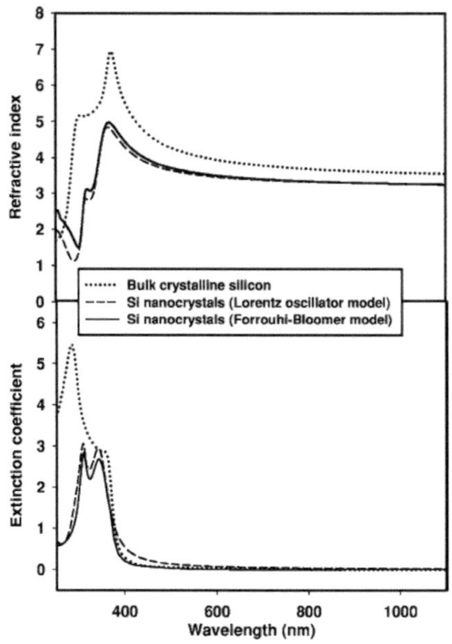


**Figure 13.** Refractive index (n) and extinction coefficient (k) of the Si NC and bulk Si as function of wavelength Ref.[79].

### 4.4. Mechanical properties

Mechanical properties of nano-materials are mainly focused on their constitutive response as well as on the fundamental physical mechanisms. Numbers of characteristics of mechanical behavior are presented to-date, like young modulus, shear modulus, bulk modulus, elastic constants and compressive strength etc. However literature shows that particularly in semiconducting NCs, these properties are yet not very much exploited; a very few reports are available on the bulk modulus of NCs which actually tells about the hardness of the material and inversely shows the compressibility of the same structure. The first-principles calculations using the LDA as exchange correlation functional, the variation in the bulk modulus of silicon nanocrystal have been investigated. On decreasing the size of the NC an enhancement in the bulk modulus has been observed [88]. In case of CdSe NCs structural transformation has been studied using high pressure X-ray diffraction and also calculates the bulk modulus ($B_0$), defined as the reciprocal of the volume compressibility, as ±74 GPa for rock salt phase of NCs and in close match with others report data for CdSe bulk and NCs. However due to the overlap of the numerous wurtzite diffraction peaks the exact value of $B_0$ has not achieved [89]. Another example is GaN where its bulk and NCs have been studied using X-ray diffraction and calculated bulk moduli are reported as 187 and 319 GPa for the wurtzite phase of bulk and NCs respectively, although the respective NaCl phases are found to have very similar bulk moduli 208 and 206 GPa [90]. A in-situ high-pressure study observes semiconductor to metal phase transitions in ZnS materials with average grain sizes of 10 μm and 11 nm. Bulk modulus and its pressure derivative ($B_0'$), of B3 phase has been estimated from the Birch-Murnaghan equation of state in a pressure range from 0 to 9 GPa as $B_0 = 72 \pm 7$ GPa and $B_0' = 9 \pm 3$ for 11 nm

ZnS and $B_0$ = 68±3 GPa and $B_0'$ = 7±1 for 10 μm ZnS NCs [91]. Another example is X-ray diffraction study of pressure-induced phase transformations in ZnO nanorods shows that the reduction of the particle size for ZnO crystallites leads to a significant increase of the bulk modulus and the pressure range of the coexistence of the wurtzite and rocksalt phases suggest that hardness of ZnO nanorods is higher than its bulk counterpart [92]. Raman study of nanocrystalline ZnO with different average crystalline sizes, predicts that the average elastic modulus of nanocrystalline ZnO shows a non-monotonic variation with crystalline size, suggests that the non-monotonic behavior is due to interplay between the elastic properties of the individual grains and those of the intergranular region [93]. A study on AlN NCs /nanowires reveals that the wurtzite structure of AlN NCs and nanowires have the $B_0$ ~215.8 GPa and ~208.8 GPa, respectively, however the $B_0$ of rocksalt phase of AlN NCs and nanowires are ~312.6 GPa, and 324.9 GPa, respectively. This shows that the decrease of particle size can appreciably lead to an increase of the $B_0$ [94].

## 4.5. Magnetic properties

Scientists divides the information technology into two parts one to exploits the degree of freedom of electronic charge in semiconductors to advances the information and other in magnetic materials where the spin degree of freedom is used to preserve the information. Spin of the charge carriers plays a slight role on semiconductor devices since most of the well known semiconductor devices are based on non-magnetic Si and GaAs which shows very less effect of spin. In contrast the superior spin related features have been acknowledged in magnetic semiconductors and diluted magnetic semiconductors (DMS) because of the coexistence of both magnetic and semiconductor properties. Nano magneto electronics is a newly developed field, where the two degrees of freedom, the charge and spin of the carriers, are utilized simultaneously to create new spintronics devices. $Mn^{2+}$ or $Co^{2+}$ the transition metal ions are the important dopants as they act as optically active impurities and localized spins in semiconductors and responsible for the optical and magnetic functionalities. Magnetic circular dichroism (MCD) spectroscopy of Mn doped and ZnS coated CdS NCs gives the electronic structures of Mn impurities and calculate the amount of Mn composition in NCs. Fig.14 shows the temperature dependence of the Zeeman splitting ($\Delta E_z$) of 0.2, 2, and 10 mol % Mn doped CdS NCs, where the MCD peak signal is allotted to Zeeman splitting of the lowest exciton. The study reveals that MCD signal directly depends on the Zeeman splitting and the first derivative of the absorption spectrum, though the Zeeman splitting is smaller than the bandwidth of the lowest exciton absorption [95].

Magnetic ion doped semiconductors are called diluted magnetic semiconductors (DMSs) [96] and these DMSs are expected to be a key material for spintronics devices due to their spin-dependent effects [97]. According to the literature not much report are available on the magnetic properties of semiconductor NCs. However, Magnetic properties of ZnS and ZnO NCs doped with various concentrations of a transition metals have been analyzed. Over a wide range of dopant concentrations without changing the size (1.6 and 4.7 nm) of the NC, the study clearly reveals the magnetic properties of such doped systems, remaining paramagnetic down to the lowest temperature, can give precise information concerning the dopant level in such samples [98,99]. ZnO NC with Co doping and without capping agent shows ferromagnetic and paramagnetic behavior, respectively. Co intra d-d transition raises in ferromagnetic NCs which may be because of the profound bound states at the NCs surface. Fig. 15 shows the variation of magnetization as a function of magnetic field for Co doped ZnO NCs and in case of washed NCs (also called as-prepared) the magnetization is directly proportional to the applied magnetic field lacking hysteresis, which shows paramagnetic behavior. However the ferromagnetic nature with hysteresis has been seen in the O capped NCs sample. As the high magnetic field increases the paramagnetic behavior is also predicted, which indicates that paramagnetic and ferromagnetic NCs coexist in these samples [100].

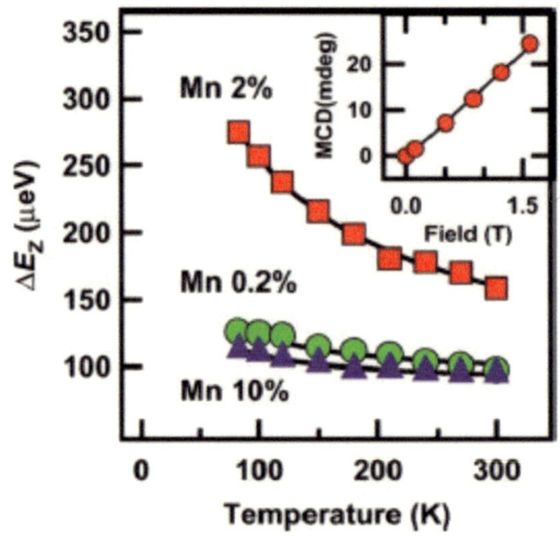


**Figure 14.** (a) Dependence of the Zeeman splitting energy on Temperature of CdS NCs doped with 0.2, 2, and 10 mol % Mn ions. The inset shows the magnetic field dependence of the MCD signal intensity in 2mol % Mn doped CdS NCs Ref.[95].

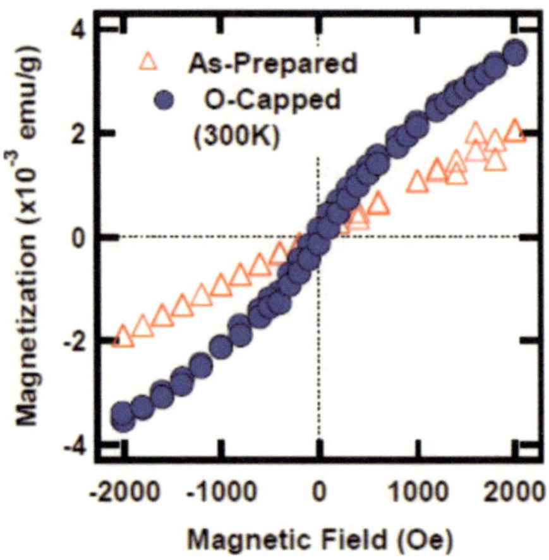


**Figure 15.** Variation of magnetization with respect to magnetic field at 300K of Co doped ZnO NCs Ref.[100].

Mn doped GaN NCs with the average diameters ranging from 4-18 nm has been prepared at low temperature and reveals that the NCs exhibit ferromagnetism with magnetization and Curie temperature values increasing with the concentration of Mn and particle size [101]. The magnetic properties of hydrogenated Si NCs doped with pairs of Mn atoms were investigated using spin density functional theory, where the two sites occupied by Mn, shows strong dependence of obtained formation energies and total magnetic moments. Pairs at sites with the same character tend to ferromagnetic spin arrangements, significantly influenced by their noncollinearity [102]. As a theoretical understanding one Ab-initio pseudopotentials approach has been employed to investigate the electronic and magnetic properties of Mn doped Ge, GaAs, and ZnSe NCs. To predict that the ferromagnetic and half-metallicity trends found in the bulk are preserved in the NCs. Due to localization of Mn states, they are less affected by quantum confinement. As a result in NCs, the Mn-related impurity states become much deeper in the gap with decreasing size [103].

## 4.6. Structural phase transition

The effects of crystal size on the stability of NC is of considerable interest because for utilizing a material in several applications, information about crystal structure at that particular temperature and pressure is must. In nano-materials the reduction in the size of the crystal changes its entire properties; this may be because of the change in the lattice parameters for a given phase and also due to pressure effects. If we assume a spherical NC, then pressure applied by the surface on the core of the NC will lead to the relaxation of the first atomic planes and therefore to an increase or a decrease in the lattice parameter. This effect is well seen in semiconductors, due to the adsorption state of the surface and surface itself reconstructs themselves to minimize the number of dangling bonds, which produces more stresses in the upper layers. Observation shows that in NC, these stresses will also contribute to the surface pressure and to the reduction or extension of atomic bonds. One obvious question arises that how will the relative stability of different possible solid structures change for NCs with respect to bulk materials? One way to answer this question is to use the pressure to force nanostructured materials to get convert it from one solid structure to another. It has been known classically that when pressure is applied the structure of the solid changes and hence its density as well as volume, that leads to the overlapping of the electron shells and structural phase transformation may take place. In III-V compound semiconductors the zinc blende to rocksalt phase transition are generally expected and intermediate phase NiAs are also seen sometimes and further the structure stabilized in a high pressure CsCl phase. The efforts on the high pressure behavior of NC starts from the year 1994 by S. H. Tolbert and A. P. Alivisatos [104] as both are the pioneer workers in the field of high pressure structural transformation in NCs. They discussed the stability of CdSe NCs in wurtzite and rocksalt phases and observed the phase transition in the range of 3.6 to 4.9 GPa for crystallites radius ranging from 21 to 10 Å. There are also some reports on the high pressure behaviour of Si and CdS NCs, shows that the smaller the crystallite higher the transition pressure, which has been explained by them in terms of surface energy differences between the phases involved [105,106]. Qadri et al. [107]

reported that the effect of reduced grain size in PbS NCs elevates the transition pressure, while the compressibility increases with decreasing grain size. An enhancement of transition pressure has been observed in ZnO NCs [108], ZnS [5] and PbS [6] in comparison to its bulk counterpart. In an another report the pressure induced I–II transition has been studied for nanocrystalline Ge of size 13, 49 and 100 nm using synchrotron x-ray diffraction, where the bulk modulus and the transition pressure increases with decrease in particle size for both Ge-I and Ge-II, but the percentage volume collapse at the transition remains constant [109]. The hexagonal AlN nanocrystals have a particle size of ~10 nm (Fig. 16) shows wurtzite to rocksalt phase transition at around 14.5 GPa, which is significantly lower than the transition pressure of 22.9 GPa observed for the bulk AlN by using the same technique. Fig.17 clearly illustrates that at the phase transition pressure of 14.5 GPa the high-pressure rocksalt phase is about 20.5% denser than the hexagonal wurtzite phase [110].

**Figure 16.** TEM image of AlN NCs Ref.[110].

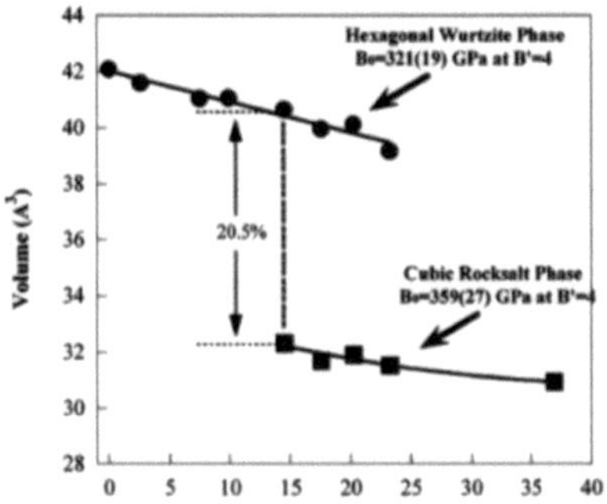


**Figure 17.** Equation of state plot and the relative volumetric variation of AlN upon the phase transition at 14.5 GPa.

Wen et al. [111] have studied the pressure induced phase transition in CdSe and ZnO NCs and found that the transition pressure increases as the nanocrystals size reduces. A pressure induced first-order structural phase transition from wurtzite to rocksalt type structure has been observed in GaN NCs at around 48.8 GPa using x-ray diffraction technique [112]. Theoretically [58] the structural stability and high pressure behavior of AlAs NCs has been analysed in various phases such as wurtzite (B4), zincblende (B3), CsCl (B2) and NaCl (B1), and observes the structural transformations from B3→B1 at around 8.9 GPa, B3→B2 at 7.12 GPa and from B3→B4 at 3.88 GPa, which is probably being the first report. The calculated values of transition pressures for AlAs NCs are lower in comparison to its bulk counterpart [57, 113,114]. The amount of volume collapse at the transition pressure has also been analyzed and the computed volume collapse for B3→B4, B3→B2 and B3→B1 transitions are found to be 5.8%, 3.5% and 1.5% respectively as illustrated in Fig.18.

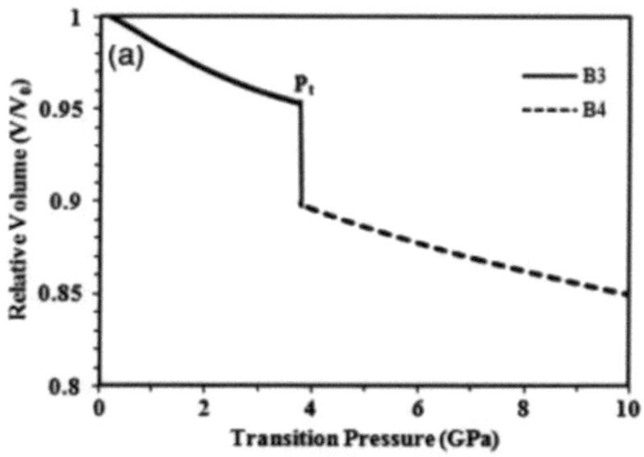


**Figure 18.** Relative volume as a function transition pressure for AlAs NCs Ref.[58].

Similar approach has been employed to investigate the structural stability of ~1nm sized AlSb NC [58] in its B3, B1 and B2 type phases under high compression and confirms that the B3 type phase is the most stable amongst the other considered phases, further under compression, the original B3 type phase of AlSb NC transforms to B1 type phase at a pressure of about 8.9 GPa, which is comparatively larger than that of bulk crystal. This discussion concludes that the transition pressure strictly depends on the size of the NC and may vary accordingly.

## 5. Conclusion

In this chapter we have described four topics; first, we have explained the fundamentals behind the peculiar properties of semiconductor NCs, the factors that makes them different from their bulk counterparts. Second, recent applications of NCs in electronic, optoelectronics, photonics as well as in medicines have been discussed. Third, we referred the present synthesis and characterization techniques for these tiny crystals, discussed various recent methods of controlled synthesis on growth and size of NCs. Fourth, we have remarked on the various properties of semiconductor NCs such as structural, electronic, optical, mechanical, magnetic properties of NCs with a detailed description of the structural transformations of semiconductor NCs under the application of pressure. We believe that the discussion on semiconductor NCs with recent results and ideas will be certainly be helpful, especially to variety of scientific community like physicists, chemists, and biologists, nano researchers/scientists for the advancements of science and technology, and its exploitation in variety of applications.

## Author details

Anurag Srivastava and Neha Tyagi
*Advance Material Research Group, Computational Nanoscience and Technology Laboratory, ABV- Indian Institute of Information Technology and Management, Gwalior (M.P.), India*

## Acknowledgement

Authors are grateful to ABV-Indian Institute of Information Technology and Management, Gwalior for providing the infrastructural support. One of us Neha Tyagi is thankful to ABV-IIITM for the award of Ph.D scholarship.

## 6. References

[1] Achermann M., Petruska M. A., Kos S., Smith D. L., Koleske D. D., Ekimov V. I. Energy-transfer pumping of semiconductor nanocrystals using an epitaxial quantumwell. Nature 2004; 429,642-646.
[2] Huynh W. U., Dittmer J. J., Alivisatos A. P. Hybrid Nanorod-Polymer Solar Cells. Science 2002; 295(5564) 2425-2427.
[3] Klein D. L., Roth R., Lim A. K. L., Alivisatos A. P., McEuen P. L. A single-electron transistor made from a cadmium selenide nanocrystal. Nature 1997;389(6652)699-701.
[4] Boinovich L.B., Emelyanenko A. M. Hydrophobic materials and coatings: principles of design, properties and applications. Russian Chemical Reviews 2008;77(7)583-600.
[5] Jiang J. Z., Gerward L., Frost D., Secco R., Peyronneacu J., Olsen J. S. Grain-size effect on pressure-induced semiconductor-to-metal transition in ZnS. Journal of Applied Physics 1999;86(11)6608-6610.
[6] Jiang J. Z., Gerward L., Frost D., Secco R., Peyronneacu J., Olsen J. S., Truckenbrodt J. Phase transformation and conductivity in nanocrystal PbS under pressure. Journal of Applied Physics 2000;87(5) 2658-2660.
[7] Shiang J.J., Kadavanich A.V., Grubbs R.K., Alivisatos A.P. Symmetry of Annealed Wurtzite CdSe Nanocrystals: Assignment to the C-3v Point Group. Journal of Physical Chemistry 1995;99(48)17417-17422.
[8] Colvin V. L., Schlamp M. C., Alivisatos A. P. Light-emitting diodes made from cadmium selenide nanocrystals and a semiconducting polymer. Nature 1994;370(6488) 354-357.
[9] Peng X., Manna L., Yang W., Wickham J., Scher E., Kadavanich A., Alivisatos A. P. Shape control of CdSe nanocrystals. Nature 2000;404(6773) 59-61.
[10] Achermann M., Petruska M. A., Koleske D. D., Crawford M. H., Klimov V. I. Nanocrystal-based light-emitting diodes utilizing high-efficiency nonradiative energy transfer for color conversion. Nano Letters 2006;6(7)1396-1400.
[11] Tessler N., Medvedev V., Kazes M., Kan S.H., Banin U. Efficient Near-Infrared Polymer Nanocrystal Light-Emitting Diodes. Science 2002;295(5559) 1506-1508.
[12] Mueller A.H., Petruska M.A., Achermann M., Werder D.J., Akhadov E.A., Koleske D.D., Hoffbauer M.A., Klimov V.I. Multicolor light-emitting diodes based on semiconductor nanocrystals encapsulated in GaN charge injection layers. Nano Letters 2005;5(6)1039-1044.

[13] Peralvarez M., Barreto J., Carreras J., Morales A., Urrios D. N., Lebour Y., Dominguez C., Garrido B. Si-nanocrystal-based LEDs fabricated by ion implantation and plasma-enhanced chemical vapour deposition. Nanotechnology 2009;20(405201)1-10.

[14] Gudiksen M.S., Maher K. N., Ouyang L., Park Hongkun. Electroluminescence from a Single-Nanocrystal Transistor, Nano Letters 2005;5(11) 2257-2261.

[15] Kahng D., Sze S. M. A floating-gate and its application to memory devices, The Bell System Technical Journal 1967;46(4) 1288-1295.

[16] Tiwari S., Rana F., Hanafi H., Hartstein A., Crabbe E.F., Chan K. A silicon nanocrystals based memory. Applied Physics Letters 1996;68(10) 1377-1379

[17] Chang T. C., Jian F. Y., Chen S. C., Tsai Y. T. Development in nanocrystal memory. Materials today 2011;14(12) 608-615.

[18] Ammendola G., Ancarani V., Triolo V., Bileci M., Corso D. , Crupi I. , Perniola L., Gerardi C., Lombardo S., DeSalvo B. Nanocrystals memories for FLASH device applications. Solid-State Electronics 2004;48 (9)1483-1488.

[19] Kim J. K., Cheong H. J., Kim Y., Yi J. Y., Bark H. J., Bang S. H. and Cho J. H. Rapid-thermal-annealing effect on lateral charge loss in metal-oxide-semiconductor capacitors with Ge nanocrystals. Applied Physics Letters 2003;82(15) 2527-2529.

[20] Lu X. B., Lee P. F., Dai J. Y. Synthesis and memory effect study of Ge nanocrystals embedded in $LaAlO_3$ high-k dielectrics. Applied Physics Letters 2005;86(203111) 1-3.

[21] Du H., Chen C., Krishnan R., Krauss T. D., Harbold J. M., Wise F. W., Thomas M. G., Silcox J. Optical Properties of Colloidal PbSe Nanocrystals. Nano Letters 2002;2(11)1321-1324.

[22] Choudhury K. R., Sahoo Y., Ohulchanskyy T. Y., Prasad P. N. Efficient photoconductive devices at infrared wavelengths using quantum dot-polymer nanocomposites. Applied Physics Letters 2005;87 (7) 073110-1 073110-3.

[23] Qi D., Fischbein M., Drndic M., Selmic S. Efficient polymer-nanocrystal quantum-dot photodetectors. Applied Physics Letters 2005;86 (9) 093103-1 093103-3.

[24] Ma X., Yuan B., Yan Z. The photodetector of Ge nanocrystals/Si for 1.55 μm operation deposited by pulsed laser deposition, Optics Communications 2006;260(1) 337-339.

[25] Guimard D., Morihara R., Bordel D., Tanabe K., Wakayama Y., Nishioka M., Arakawa Y. Fabrication of InAs/GaAs quantum dot solar cells with enhanced photocurrent and without degradation of open circuit voltage. Applied Physics Letters 2010;96 (20) 203507-1 203507-3.

[26] Gur I., Fromer N.A., Geier M. L., Alivisatos A. P. Air-Stable All-Inorganic Nanocrystal Solar Cells Processed from Solution. Science 2005;310(5747) 462-465.

[27] Gross N. Y., Harari M. S., Zimin M., Kababya S., Schmidt A., Tessler N. Molecular control of quantum-dot internal electric field and its application to CdSe-based solar cells. Nature Materials 2011;10(12) 974-979.

[28] Svrcek V., Mariotti D., Nagai T., Shibata Y., Turkevych I., Kondo M. Photovoltaic Applications of Silicon Nanocrystal Based Nanostructures Induced by Nanosecond Laser Fragmentation in Liquid Media. Journal of Physical Chemistry C 2011;115(12)5084-5093.

[29] Katteboinaa Suman, V S R Chandrasekhar. P, Balaji S. Drug Nanocrystals: a Novel Formulation Approach for Poorly Soluble Drugs. International Journal of PharmTech Research 2009;1(3)682-694 (2009).

[30] Liversidge E. M. M., Liversidge G. G. Drug Nanoparticles: Formulating Poorly Water-Soluble Compounds. Toxicologic Pathology 2008;36(1)43-48.

[31] Tran P.T., Goldman E.R., Anderson G.P., Mauro J.M., Mattoussi H. Use of luminescent CdSe-ZnS nanocrystal bioconjugates in quantum dot-based nanosensors. Physica Status Solidi B 2002;229(1), 427-432.

[32] Dubertret B., Skourides P., Norris D. J., Noireaux V., Brivanlou A. H., Libchaber A. In vivo imagining of quantum dots encapsulated in phospholipid micelles. Science 2002;298(5599) 1759-1762.

[33] Parak W. J., Boudreau R., Gros M. L., Gerion D., Zanchet D., Micheel C. M., Williams S. C., Alivisatos A. P., Larabell C. Cell motility and metastatic potential studies based on quantum dot imaging of phagokinetic tracks. Advanced Materials 2002;14(12) 882-885.

[34] Taylor J. R., Fang M. M., Nie S. M. Probing specific sequences on single DNA molecules with bioconjugated fluorescent nanoparticales. Analytical Chemistry 2000;72(9)1979-1986.

[35] Parak W. J., Pellegrino T., Plank C. Labelling of cells with quantum dots. Nanotechnology 2005;16(2) R9-R25.

[36] Brus L. Electronic wave functions in semiconductor clusters: experiment and theory. Journal of Physical Chemistry 1986;90(12)2555-2560.

[37] Steigerwald M. L., Alivisatos A. P., Gibson J. M., Harris T. D., Kortan R., Muller A. J., Thayer A. M., Duncan T. M., Douglass D. C. and Brus L. E. Surface derivatization and isolation of semiconductor cluster molecules. Journal of American Chemical Society 1988;110(10)3046-3050.

[38] Yong K.-T., Sahoo Y., Swihart M. T., Prasad P. N. Shape Control of CdS Nanocrystals in One-Pot Synthesis. Journal of Physical Chemistry C 2007;111(6) 2447-2458.

[39] Wang W., Poudel B., Huang J. Y., Wang D. Z., Kunwar S., Ren Z. F. Synthesis of gram-scale germanium nanocrystals by a low-temperature inverse micelle solvothermal route. Nanotechnology 2005;16(8)1126-1129.

[40] Chen L., Holmes J. D., Garcia S. R., Morris M. A. Facile Synthesis of Monodisperse ZnO Nanocrystals by Direct Liquid Phase Precipitation. Journal of Nanomaterials 2011; 2011, 1-9.

[41] Yu M., Hao X., Cui D., Wang Q., Xu X., Jiang M. Synthesis of aluminium nitride nanocrystals and their catalytic effect on the polymerization of benzene. Nanotechnology 2003;14(1) 29.

[42] Yang H., Yao X., Huang D. Sol gel synthesis and photoluminescence of AlP nanocrystals embedded in silica glasses. Optical Materials 2007;29(7)747-752.

[43] Yang H.Q., Yao X., Huang D.M., Wang X. J., Shi H. Z., Zhang B. L., Liu S. X., Fang Y. Sol-gel synthesis and photoluminescence of III-V semiconductor InAs nanocrystals embedded in silica glasses. Journal of Nanoscience and Nanotechnology 2005;5(5)786-9.

[44] Yang H., Huang D., Wang X., Gu X., Wang Fujian, Xie S., Yao X. Sol-Gel Synthesis of Luminesent InP Nanocrystals Embedded in Silica Glasses. Journal of Nanoscience and Nanotechnology 2005;5 (10)1737-1740.

[45] Nayak J., Mythili R., Vijayalakshmi M., Sahu S. N. Size quantization effect in GaAs nanocrystals. Physica E: Low-dimensional Systems and Nanostructures 2004; 24(3-4)227-233.

[46] Yuk J. M., Kim K., Lee Z., Watanabe M., Zettl A., Kim T. W., No Y. S., Choi W. K., Lee J. Y. Direct fabrication of zero- and one-dimensional metal nanocrystals by thermally assisted electromigration. ACS Nano 2010;4(6)2999-3004.

[47] Chiu C. W., Liao T. W., Tsai K. Y., Wang F. M., Suen Y. W., Kuan C. H. Fabrication method of high-quality Ge nanocrystals on patterned Si substrates by local melting point control. Nanotechnology 2011;22(27)275604, 1-5.

[48] Namatsu H., Kurihara K., Nagase M., Makino T. Fabrication of 2-nm-wide silicon quantum wires through a combination of a partially-shifted resist pattern and orientation-dependent etching. Applied Physics Letters 1997;70(5) 619-621.

[49] Murray C. B., Norris D. J., Bawendi M. G. Synthesis and characterization of nearly monodisperse CdE (E = sulfur, selenium, tellurium) semiconductor nanocrystallites. Journal of American Chemical Society 1993;115(19) 8706-8715.

[50] McCaldin J. O. Current approaches to p-n junction in wider band gap II-VI semiconductors. Journal of Vaccum Science and Technology A 1990;8(2), 1188-1193.

[51] Alphandery E., Nicholas R. J., Mason N. J., Lyapin S. G., Klipstein P. C. Photoluminescence of self-assembled InSb quantum dots grown on GaSb as a function of excitation power, temperature, and magnetic field. Physical Review B 2002;65,115322-1 115322-7.

[52] Xie Y., Qian Y.T., Wang W.Z., Zhang S.Y., Zhang Y.H. A Benzene-Thermal Synthetic Route to Nanocrystalline GaN. Science 1996;272(5270)1926-1927.

[53] Dvorak M.D., Justus B.L., Gaskill D. K., Hendershot D.G. Nonlinear absorption and refraction of quantum confined InP nanocrystals grown in porous glass. Applied Physics Letters 1995;66(7) 804-806.

[54] Kim S. S., Bang K.I., Kwak J., Lim K. S. Growth of Silicon Nanocrystals by Low-Temperature Photo Chemical Vapor Deposition. Japanese Journal of Applied Physics 2006;45(1)L46-L49.

[55] Costales A., Blanco M. A., Francisco E., Solano C. J. F., Pendas A. M. Theoretical Simulation of AlN Nanocrystals. Journal of Physical Chemistry C 2008;112(17) 6667-6676.

[56] Franceschetti A. Structural and electronic properties of PbSe nanocrystals from first principles. Physical Review B 2008;78,075418-1 075418-6.

[57] Srivastava A., Tyagi N., Singh R.K. Pressure induced phase transformation and electronic properties of AlAs. Materials Chemistry Physics 2011;125(1-2)66-71.

[58] Srivastava A., Tyagi N. High pressure behavior of AlAs nanocrystals: the first-principle study. High Pressure Research iFirst 2012, 1-5.

[59] Tyagi N., Srivastava A. Structural Phase Transition and Electronic Properties of AlSb Nanocrystal. Proc. Intl. Conf. on Nanosci. Eng. Tech. (ICONSET 2011) 978-1-4673-0073-5/11/$26.00 @2011 IEEE IEEEXplore 2011, 421-423.

[60] Tyagi N., Srivastava A. Electronic properties of AlN nanocrystal: A first principle study. XVI National Seminar on Ferroelectrics and Dielectrics and (NSFD-XVI), American Institute of Physics Conference Proc. 2011;1372, 252-262.

[61] Z.-G., Bai Y.-J., Cui D.-L., Hao X.-P., Wang L.-M., Wang Q.-L., Xu X.-G. Increase of lattice constant in GaP nanocrystals. Journal of Crystal Growth 2002;242(3-4) 486–490.

[62] Efros Al. L., Efros A. L. Interband absorption of light in a semiconductor sphere. Soviet Physics Semiconductors 1982;16(7) 772-775.

[63] Buda F., Kohanoff J., Parrinello M. Optical properties of porous silicon: A first-principle study. Physical Review Letters 1992;69(8) 1272-1275.

[64] Krishna M. V. R., Friesner R. A. Exciton spectra of semiconductor clusters. Physical Review Letters 1991;67(5) 629-632.

[65] Jancu J. M., Scholz R., Beltram F., Bassani F. Empirical spds* tight-binding calculations for cubic semiconductors: General method and materials parameters. Physical Review B 1998;57(11)6493-6507.

[66] Einevoll G.T. Confinement of excitons in quantum dots. Physical Review B 1992;45(7) 3410-3417.

[67] Nair S.V., Ramaniah L.M., Rustagi K.C. Electron states in a quantum dot in an effective-bond-orbital model. Physical Review B 1992;45(11) 5969-5979.

[68] Buuren T. V., Dinh L.N., Chase L.L., Siekhaus W.J., Terminello L.J. Changes in the Electronic Properties of Si Nanocrystals as a Function of Particle Size. Physical Review Letters 1998;80(17) 3803-3806.

[69] Wang L. W., Zunger A. Local-density-derived semiempirical pseudopotentials. Physical Review B 1995;51(24) 17398-17416.

[70] Fu H., Zunger A. Local-density-derived semiempirical nonlocal pseudopotentials for InP with applications to large quantum dots. Physical Review B 1997;55(3)1642-1653.

[71] Lippens P.E., Lannoo M. Calculation of the band gap for small CdS and ZnS crystallites. Physical Review B 1989;39(15)10935-10942.

[72] Sapra S., Viswanatha R., Sarma D. D. An accurate description of quantum size effects in InP nanocrystallites over a wide range of sizes. Journal of Physics D: Applied Physics 2003;36(13)1595-9.

[73] Sapra S., Shanthi N., Sarma D. D. Realistic tight-binding model for the electronic structure of II-VI semiconductors. Physical Review B 2002;66(8)205202-1 205202-8.

[74] Diaz J. G., Bryant G. W. Electronic and optical fine structure of GaAs nanocrystals: The role of d orbitals in a tight-binding approach. Physical Review B 2006;73(7) 075329-1 075329-9.

[75] Mizel Ari, Cohen M. L. Electronic energy levels in semiconductor nanocrystals: A Wannier function approach. Physical Review B 1997;56(11)6737-6741.

[76] Beard M. C., Knutsen K. P., Yu P., Luther J. M., Song Q., Metzger W. K., Ellingson R. J., Nozik A. J. Multiple exciton generation in colloidal silicon nanocrystals. Nano Letters 2007;7(8)2506-2512.

[77] Schaller R. D., Pietryga J. M., Goupalov S. V., Petruska M. A., Ivanov S. A., Klimov V. I. Breaking the Phonon Bottleneck in Semiconductor Nanocrystals via Multiphonon Emission Induced by Intrinsic Nonadiabatic Interactions. Physical Review Letters 2005;95(19)196401-1 196401-4.

[78] Cullis A. G., Canham L. T., Calcott P. D. J. The structural and luminescence properties of porous silicon. Journal of Applied Physics 1997;82(3)909-965.

[79] Ding L., Chen T. P., Liu Y., Ng C. Y., Fung S. Optical properties of silicon nanocrystals embedded in a $SiO_2$ matrix. Physical Review B 2005;72(12)125419,1-7.

[80] Kanemitsu Y., Tanaka H., Kushida T., Min K.S., Atwater H. A. Luminescence properties of GaAs nanocrystals fabricated by sequential ion implantation. Journal of Luminesence 2000;87-89,432-434.

[81] Nirmal M., Dabbousi B. O., Bawendi M. G., Macklin J. J., Trautman J. K., Harris T. D., Brus L. E. Fluorescence intermittency in single cadmium selenide nanocrystals. Nature 1996;383(6603)802-804.

[82] Schlamp M. C., Peng X., Alivisatos A. P. Improved efficiencies in light emitting diodes made with CdSe(CdS) core/shell type nanocrystals and a semiconducting polymer. Journal of Applied Physics 1997;82(11)5837-5842.

[83] Jiang X., Schaller R. D., Lee S. B., Pietryga J. M., Klimov V. I., Zakhidov A. A. PbSe nanocrystal/conducting polymer solar cells with an infrared response to 2 micron, Journal Materials Research 2007;22(8) 2204-2210.

[84] Schaller R. D., Klimov V. I. High Efficiency Carrier Multiplication in PbSe Nanocrystals: Implications for Solar Energy Conversion. Physical Review Letters 2004;92(18)186601,1-4.

[85] Green M. A. Third generation photovoltaics: solar cells for 2020 and beyond. Physica E 2002;14(1-2) 65-70.

[86] Nozik A. J. Spectroscopy and hot electron relaxation dynamics in semiconductor quantumwells and quantum dots. Annual Review of Physical Chemistry 2001;52, 193-231.

[87] Zhou Z., Brus L., Friesner R. Electronic Structure and Luminescence of 1.1- and 1.4-nm Silicon Nanocrystals: Oxide Shell versus Hydrogen Passivation. Nano Letters 2003;3(2)163-167.

[88] Cherian R., Gerard C., Mahadevan P., Cuong N. T., Maezono R. Size dependence of the bulk modulus of semiconductor nanocrystals from first-principles calculations. Physical Review B 2010;82(23) 235321, 1-7.

[89] Tolbert S. H., Alivisatos A. P. The wurtzite to rock salt structural transformation in CdSe nanocrystals under high pressure. Journal of Chemical Physics 1995;102(11)4642-4656.

[90] Jorgensen J.-E., Jakobsen J. M., Jiang J. Z., Gerward L., Olsen J. S. High-pressure X-ray diffraction study of bulk- and nanocrystalline GaN. Journal of Applied Crystallography 2003;36(2) 920-925.

[91] Jiang J. Z. Phase transformations in nanocrystals. Journal of Materials Science 2004;39(16-17)5103- 5110.

[92] Wu X., Wu Z., Guoc L., Liu C., Liu J., Li X., Xu H. Pressure-induced phase transformation in controlled shape ZnO nanorods. Solid State Communications 2005;135(11-12)780-784.

[93] Panchal V., Ghosh S., Gohil S., Kulkarni N., Ayyub P. Non-monotonic size dependence of the elastic modulus of nanocrystalline ZnO embedded in a nanocrystalline silver matrix. Journal of Physics: Condensed Matter 2008;20(34)345224,1-4.

[94] Lei W. W., Liu D., Zhang J., Cui Q. L., Zou G. T. Comparative studies of structural transition between AIN nanocrystals and nanowires. Journal of Physics: Conference Series 2008;121,162006,1-8.

[95] Taguchi S., Ishizumi A., Tayagaki T., Kanemitsu Y. Mn-Mn couplings in Mn-doped CdS nanocrystals studied by magnetic circular dichroism spectroscopy. Applied Physics Letters 2009;94(17)173101-173103.

[96] Ando K., Takahashi K., Okuda T., Umehara M. Magnetic circular dichroism of zinc-blende-phase MnTe. Physical Review B 1992;46(19)12289-12297.

[97] Furdyna J. K. Diluted magnetis semiconductor. Journal of Applied Physics 1988;64(4) R29-R64.

[98] Wolf S. A., Awschalom A. A., Buhrman R. A., Daughton J. M., Molnar S., Roukes M. L., Chtchelkanova A. Y., Treger D. M. Spintronics: A Spin-Based Electronics Vision for the Future. Science 2001;294(5546)1488-1495.

[99] Sarma D.D., Viswanatha R., Sapra S., Prakash A., Hernandez M. G. Magnetic Properties of Doped II–VI Semiconductor Nanocrystals. Journal of Nanoscience and Nanotechnology 2005;5(9)1503-1508.

[100] Taguchi S., Tayagaki T., Kanemitsu Y. Luminescence and magnetic properties of Co doped ZnO nanocrystals. IOP Conf. Series: Materials Science and Engineering 2009;6, 012029, 1-4.

[101] Biswas K., Sardar K.,Rao C. N. R. Ferromagnetism in Mn-doped GaN nanocrystals prepared solvothermally at low temperatures. Applied Physics Letters 2006;89(13) 132503-132505.

[102] Panse C., Leitsmann R., Bechstedt F. Magnetic interaction in pairwise Mn-doped Si nanocrystals. Physical Review B 2010;82(12)125205, 1-9.

[103] Huang X., Makmal A., Chelikowsky J. R., Kronik L. Size dependent spintronics properties of diluted magnetic semiconductors nanocrystals. Physical Review Letters 2005;94(23)236801, 1-4.

[104] Tolbert S. H., Alivisatos A. P. Size Dependence of a First Order Solid-Solid Phase Transition: The Wurtzite to Rock Salt Transformation in CdSe Nanocrystals. Science 1994;265(5170)373-376.

[105] Tolbert S. H., Alivisatos A.P. The Wurtzite to Rock-Salt Structural Transformation in Cdse Nanocrystals under High-Pressure. Journal of Chemical Physics 1995;102(11)4642-4656.

[106] Tolbert S. H., Herhold A. B., Brus L. E., Alivisatos A. P. Pressure-induced structural transformations in Si nanocrystals: Surface and shape effects. Physical Review Letters 1996;76(23)4384-4387.

[107] Qadri S . B., Yang J ., Ratna B. R., Skelton E. F., Hu J . Z. Pressure induced structural transitions in nanometer size particles of PbS. Applied Physics Letters 1996;69(15) 2205-2207.

[108] Jiang J . Z., Olsen J . S ., Gerward L., Frost D., Rubie D., Peyronneau J . Structural stability in nanocrystalline ZnO. Europhysics Letters 2000;50(1)48-53.

[109] Wang H., Liu J. F., He Y., Wang Y., Chen W., Jiang J. Z., Olsen J. S., Gerward L. High-pressure structural behaviour of nanocrystalline Ge. Journal of Physics: Condensed Matter 2007;19(15)156217.

[110] Wang Z., Tait K., Zhao Y., Schiferl D., Zha C., Uchida H., Downs R. T. Size-Induced Reduction of Transition Pressure and Enhancement of Bulk Modulus of AlN Nanocrystals. Journal of Physical Chemistry B 2004;108(31)11506-11508.

[111] Li S., Wen Z., Jiang Q. Pressure-induced phase transition of CdSe and ZnO nanocrystals. Scripta Materialia 2008;59(5) 526-529.

[112] Cui Q., Pan Y., Zhang W., Wang X., Zhang J., Cui T., Xie Y., Liu J., Zou G. Pressure-induced phase transition in GaN nanocrystals. Journal of Physics: Condensed Matter 2002;14(44)11041-11044.

[113] Mujica A., Needs R.J., Munoz A. First-principles pseudopotential study of the phase stability of the III-V semiconductors GaAs and AlAs. Physical Review B 1995;52(12)8881-8892.

[114] Cai J., Chen N. Theoretical study of pressure-induced phase transition in AlAs: From zinc-blende to NiAs structure. Physical Review B 2007;75(17)174116,1-8.

Chapter 8

# Surface Modification of CdSe and CdS Quantum Dots-Experimental and Density Function Theory Investigation

Liang-Yih Chen, Hung-Lung Chou, Ching-Hsiang Chen
and Chia-Hung Tseng

Additional information is available at the end of the chapter

http://dx.doi.org/10.5772/46516

## 1. Introduction

Semiconductor nanocrystals (NCs), also referred to as semiconductor quantum dots (QDs), which are small compared to the bulk exciton radius have unique properties associated with the spatial confinement of the electronic excitations. These semiconductor QDs have discrete electronic states, in contrast to the bulk band structure, with an effective band gap blue shifted from that of the bulk. Due to their unique properties, QDs have been of great interest for fundamental research and industrial development in recent years [1-2]. The size-dependent optical properties of QDs have been actively studied during the pass decade. The synthesis and subsequent functionality of QDs for a variety of applications include photostable luminescent biological labels [3-8], light harvesters in photovoltaic devices [9-15] and as the emissive material in light-emitting devices (LEDs) [16-23]. They are also characterized by large surface to volume ratios. Unlike quantum wells and wires, experimental studies suggest that the surfaces of QDs may play a crucial role in their electronic and optical properties. However, these semiconductor QDs with large surface to volume ratios are metastable species in comparison to the corresponding bulk crystal and must be kinetically stabilized. The most common method to maintain their stability is by chemically attaching a monolayer of organic molecules to the atoms on the surfaces of QDs. These organic molecules are often called surfactants, capping groups, or ligands. In addition to the protection function, this monolayer of ligands on the surfaces of QDs provides the necessary chemical accessibility for the QDs by varying the terminal groups of the ligands pointing to the outside environment. For example, the QDs covered with hydrophobic ligands cannot be used directly in applications that require aqueous solubility or an effective charge transport property.

For both the protection and solubility function, the photoluminescence quantum yield (PL-QY) is another crucial factor for semiconductor QDs. The QDs have a large fraction of surface atoms because of their small volume; therefore, several inhomogeneous defect points occur on the large surface area. The processes that determine the luminescence QYs in semiconductor QDs have been investigated for several years. In this topic, the surface chemistry plays a crucial role in the manipulation of semiconductor QDs because it determines the dispersion interactions of the QDs in the medium, and high quantum yields and long term photostability can be achieved only through an improved understanding of surface recombination processes. Therefore, proper passivation of the surfaces of QDs is necessary to achieve a high PL-QY.

QDs of II-VI semiconductors have been extensively studied during the past two decades [24-30]. To date, the light-emitting core part of most QDs is cadmium selenide (CdSe), which can be prepared under mild conditions using well-known precursors. However, CdSe QDs have a spectral limitation at emission wavelengths shorter than 490 nm. To achieve high quantum efficiency in the blue region, cadmium sulfide (CdS) is a suitable candidate that can be prepared under mild conditions. Because the bulk CdS has an energy band gap of approximately 2.5 eV, it is easier to enable CdS QDs to emit blue light than the same sized CdSe. In connection with the improvement of the emission efficiency of CdSe and CdS QDs, several studies focused on the capping ligands introduced to the surfaces of CdS and CdSe QDs to study the variation of PL-QY [31-35]. In general, the usual method for surface modification of CdS QDs is to cap the synthesized CdS QDs with thiolate ligands during the growth period [36]. Uchihara et al. investigated the pH dependent photostability of thioglycerol-capped CdS (TG-CdS) and mercaptoacetate-capped CdS (MA-CdS) in colloidal solutions under stationary irradiation [37]. Because of the various charge properties of TG and MA, the carboxyl group of MA is neutralized by the addition of proton in the lower pH region, and the photostability decreases by lowering the pH of the solutions for MA-CdS; however, the photostability of TG-CdS are slightly influenced by the pH of the solution. Thangadurai et al. used 1.4-dithiothreitol (DTT), 2-mercaptoethanol (ME), cysteine (Cys), methionine (Meth), and glutathione (GSH) as ligands to cap the surfaces of CdS QDs to study the photo-initiated surface degradation [33]. It is noteworthy that the band edge emission of DDT capped CdS shifted to a higher energy, and this shift was in conformity with the lowest grain size. In addition, the intensity of the broadband related to the surface defect states of CdS QDs exhibited a reduced trend compared to the other samples. The surface coating with suitable thiol molecules can yield a lower grain size in the cubic phase and obtain excellent fluorescence properties with efficient quenching of the surface traps.

In contrast to the photochemical stability of thiol capped-CdS QDs, the photochemical instability of CdSe QDs capped with thiol molecules was reported by Peng et al. [34]. Based on their research, they proposed that the photochemical instability of CdSe QDs capped with thiol molecules included three distinguishable processes, as follows: (1) the photocatalytic oxidation of the thiol molecules on the surfaces of CdSe QDs; (2) the photooxidation of CdSe QDs; and (3) the precipitation of CdSe QDs. Thiols are the most widely used ligands for stabilizing semiconductors [38-39]. However, the stability of the thiol-stabilized CdSe QDs is not satisfactory because of the photooxidation of the QDs-ligand complex using CdSe QDs as the photocatalysts. It is difficult to reproducibly apply chemical and biochemical procedures to these QDs because of their unstable nature. In addition to the use of thiol molecules to cap the surfaces of CdSe QDs, most researchers used various amines as ligands to modify the optical properties of CdSe QDs [40-44]. Talapin et al. synthesized CdSe QDs in a three-component hexadecylamine (HDA)-trioctylphosphine oxide (TOPO)-trioctylphosphine (TOP) mixture [35]. The room temperature PL-QY of as-synthesized CdSe QDs without adding HDA was in the range of 10-25%. However, the PL-QY of CdSe QDs can be improved substantially by surface passivation with HDA molecules. This indicates that PL efficiency losses are caused by insufficient passivation of the surface traps. Murray et al. examined the effect of the surface modification of CdSe QDs on the optical properties [45]. They interpreted the substantial increase in PL intensity following the addition of HDA molecules as the elimination of nonradiative decay pathways. Similar results were reported by Bullen and Mulvaney with primary, secondary, and tertiary amine [41].

Two types of alkylamine (*n*-butylamine (*n*-BA) and *n*-hexylamine (*n*-HA)) and oleic acid (OA) were used to modify the surfaces of the CdS and CdSe nanocrystals. To understand the changes of optical properties of un-modified QDs and surface modified QDs, PL spectra and PL-QY were used to characterize the emission peak position and emission efficiency after surface modification of these ligands for CdSe and CdS QDs. The PL decay kinetics for these ligand capped-QDs systems were followed by time-resolved photoluminescence (TRPL), and the spectra were analyzed in regard to a biexponential model to identify two lifetime values, that is, shorter-lifetime ($\tau_S$) and longer-lifetime ($\tau_L$). The detailed mechanism was studied by density function theory (DFT) simulation to demonstrate the binding energy and charge analyses of CdS or CdSe QDs with *n*-BA, *n*-HA, and OA.

## 2. The surface modification of CdS and CdSe QDs via organic ligands

The capping of CdS and CdSe QDs consisting of semiconductor cores surrounded by organic ligands has attracted considerable interest for applications in materials science and nanotechnology [46-48]. Although these studies enabled syntheses of stable capping semiconductor QDs with various sizes, shapes, and compositions, few studies have been conducted on the surface structures and properties of capping ligands. Information regarding the nature and chemical properties of the binding between QDs and their ligands is limited. However, compared to atoms on the flat surface of bulk substrates, the binding abilities of atoms on curved surfaces may be affected by their diverse structural environments and size-dependent electron configuration. In this study, the synthesis of CdS and CdSe QDs was conducted in a noncoordinating solvent using 1-ODE. The CdO (0.16 mmol) was mixed with 0.7 mmol OA and 4.8 g of 1-ODE in a 25 mL three-neck flask. The mixture was heated to 300 °C under Ar flow for 30 min, and subsequently injected with Se stock solution (0.1 mmol of S or Se powder dissolved in 0.62 mmol of TBP and 1 g of ODE). The solution mixture was cooled, and the nanocrystals were allowed to grow at 260 °C to reach the desired size, as determined by UV-visible absorption. To monitor the growth of QDs, a small amount of the sample (approximately 0.2 mL) was obtained through a syringe and diluted to exhibit an optical density between 0.1 and 0.2 by the addition of anhydrous toluene. The resulting CdS and CdSe QDs were suspended in toluene, and the unreacted starting materials and side products were removed by extraction and precipitation procedures. Size sorting was not performed in any of the samples. An aliquot of CdS or CdSe QDs solution was diluted with toluene to yield an optical density of approximately 0.1 at a wavelength of 350 nm (the excitation wavelength of PL). A 3 mL portion of the nanocrystals solution was mixed with various capping molecules at a fixed concentration of 5 mM for surface modification. The solution mixture was stirred in the dark at room temperature for 1 h. The CdS and CdSe QDs were subsequently precipitated with methanol and re-dispersed in toluene for characterization of the change of PL quantum yield by UV/vis absorption and photoluminescence spectroscopy.

Figure 1 shows the transmission electron microscopic (TEM) images of the as-grown CdSe and CdS QDs synthesized by the non-coordinate method. It shows that the material has a uniform size distribution and regular shape with 5 nm and formed close-packed arrays.

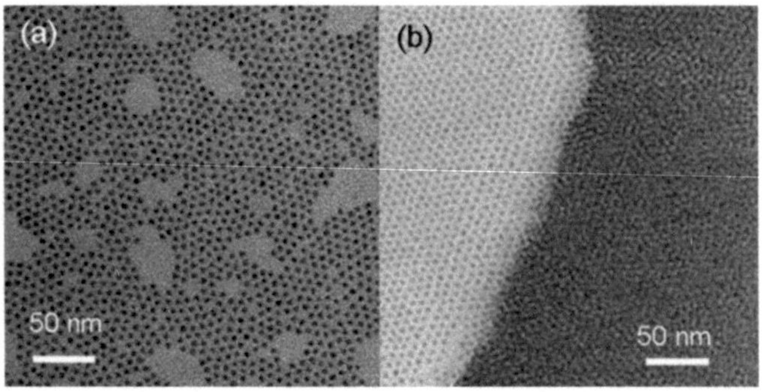


**Figure 1.** Transmission electron microscopic images of samples of (a) CdSe QDs; (b) CdS QDs.

The absorption and PL spectra of the CdSe QDs varying with growth time are shown in Figure 2. The luminescence spectra from these CdSe QDs are symmetric and narrow. However, the PL quantum yields decreased in conjunction with the growth time to 5%, as shown in Figure 2(c). The absorption and PL spectra of CdS QDs recorded for samples grown at various times are shown in Figure 3. The PL-QY increased rapidly in conjunction with the growth time initially and reached a steady value of approximately 60% after 200 s. The increasing PL-QY with crystal growth for CdS QDs exhibited contrasting behavior to several other QDs, such as CdSe, which exhibited a decreasing PL QY with crystal growth time [49]. A characteristic peak for CdS QDs was not observed in the UV and PL spectra in the initial 60 s after S precursor injection to the $Cd^{2+}$ solution. The UV absorption spectrum and the narrow emission peak with full-width at half maximum (approximately 21 nm) were observed after 60 s reaction. This behavior can be attributed to the slow growth of CdS QDs in the initial period, and their size was too small in this time domain for identification by the spectroscopy of UV-visible absorption and PL emission. Therefore, because of the slow growth process in the formation of CdS QDs, high quality QDs with large quantities of radiative surface-states with low nonradiative surface quenching defects can be obtained with the increase in PL-QY in conjunction with the growth time.

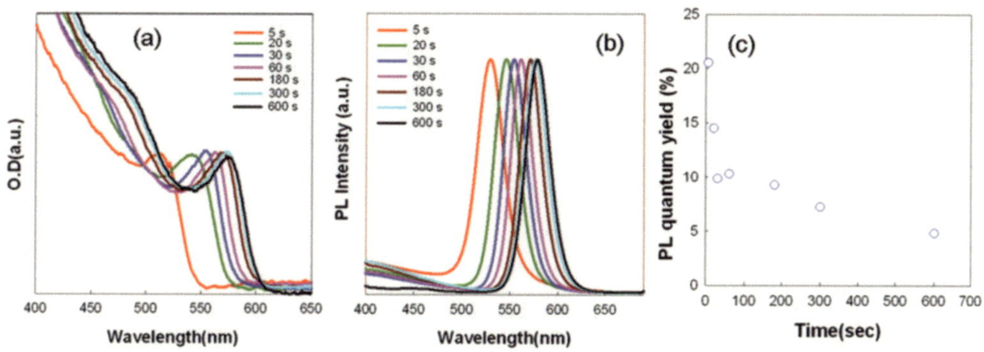


**Figure 2.** Temporal evolution of (a) UV-vis; (b) PL spectra and (c) PL quantum yield of a growth reaction of CdSe QDs.

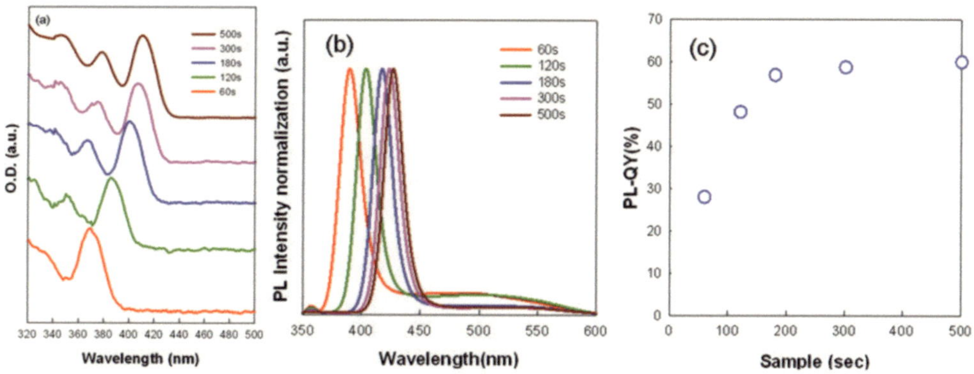


**Figure 3.** Temporal evolution of (a) UV-visible spectra; (b) PL spectra and (c) PL quantum yield of a growth reaction of CdS QDs.

Based on these observations, the control of the surface, probably a reconstructed surface of QDs, may be crucial for controlling and improving their PL properties. To understand the changes in the optical properties of the CdSe and CdS QDs upon modification by the capping ligands, *n*-BA, *n*-HA ,and OA were added to a solution of as-grown CdSe and CdS QDs solution, and the PL and the UV/vis spectra were recorded before and after the addition. The PL spectra and the quantum yields of the CdSe QDs modified by *n*-BA, *n*-HA, and OA are shown in Figure 4. As shown in Fig. 4, the positions of the luminescence emission peaks of the ligand-modified CdSe QDs shifted to lower wavelengths for the three capping molecules compared to the as-grown CdSe QDs. Moreover, Figure 4(b) shows that the PL quantum yield increased to 45% and 61% for the amines *n*-BA and *n*-HA capping CdSe QDs, respectively, whereas OA capping of CdSe QDs exhibited a decrease to 5%. For ligand capping of CdS QDs, the PL spectra and the QYs of the CdS QDs modified by *n*-BA, *n*-HA, and OA are shown in Figure 5. As shown in Figure 5(a), the PL intensity was decreased substantially by the additives in the entire wavelength region, whereas the PL spectral maximum occurred at the same wavelength for all systems. Figure 5(b) shows that the PL QY of CdS QDs decreased from 60% to 6% and 3% for *n*-BA and *n*-HA, respectively, whereas OA exhibited a decrease to 2%. Compared to the case of CdSe QDs, the three organic additives in this study exhibited contrasting behavior for the CdS QDs system.

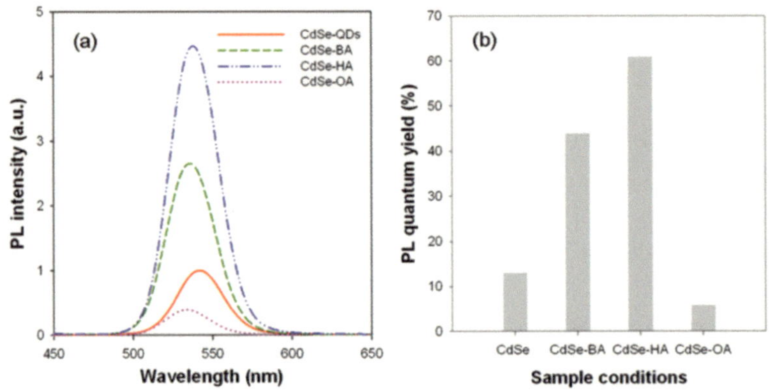


**Figure 4.** (a) Room temperature PL spectra of as-grown, *n*-BA, *n*-HA and OA modified CdSe QDs; (b) The corresponding PL quantum yield variety of CdSe QDs after capping ligand modification.

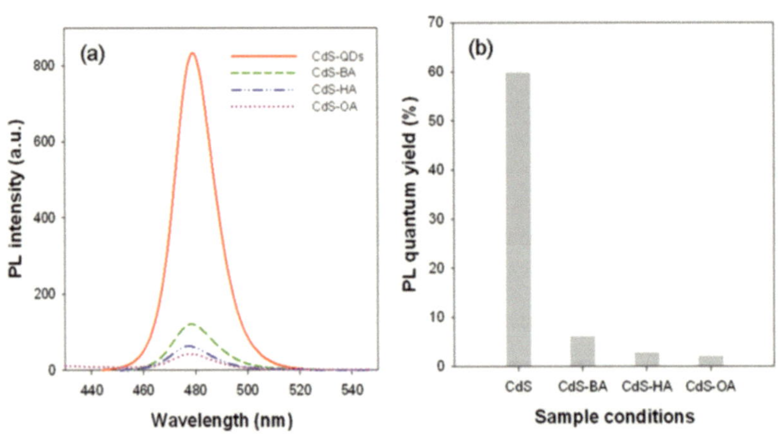


**Figure 5.** (a) Room temperature PL spectra of as-grown, *n*-BA, *n*-HA and OA capping molecule modified CdS QDs; (b) PL quantum yield of CdS QDs versus capping ligands.

Generally, it is believed that the capping ligands effectively passivate the surface states and suppress the non-radiative recombination at surface vacancies, leading to enhanced PL quantum yield. However, among the three capping agents in the case of CdSe QDs, the OA ligand exhibited contrasting behavior. In addition, the three ligands were used to quench the emission properties of CdS QDs. Generally, the relaxation process of QDs is radiative recombination. Otherwise, competing radiation-less relaxation processes are used, including carrier trapping at QD defects, charge transfer between QDs and ligand-based orbitals, and inter-QDs energy transfer. The observations in this study clearly indicate that the passivation effect of the capping ligands for CdSe and CdS QDs are more complex, indicating the requirement for a careful examination of the photo-induced charge transfer between CdSe QDs and the capping ligands.

To resolve the various behaviors of CdS QDs, time-resolved photoluminescence (TRPL) was used to probe the decay kinetics of the exciton emission of bare QDs and ligand capping QDs by *n*-BA, *n*-HA, and OA molecules. Because of the high sensitivity of TRPL for ensemble and single particle PL analysis, it is often used to determine the transient population of one or more radiative excited states [50].

## 3. The study of time resolved photoluminescence technique on surface modification of CdS and CdSe QDs.

For TRPL analysis, this study used a system with a single picosecond diode laser driver with a 375cnm laser head (integrated collimator and TE cooler for temperature stabilization was integrated by Protrustech Co., Ltd). An Andor iDus CCD with 1024 × 128 pixels was used to obtain the PL signal and the Pico Quant PMT Detector head with 200–820 nm and <250 ps IRF was integrated to obtain the TRPL signal. A theoretical model with multi-exponential analysis is often used to determine the lifetime of QDs by PL decay data. Typically, the feature of the decay of the PL intensity for QDs is a universal occurrence of a biexponential time distribution in the radiative lifetime, as shown in the following equation,

$$I(t) = A_S e^{-t/\tau_S} + A_L e^{-t/\tau_L} \tag{1}$$

where $\tau_S$ and $\tau_L$ represent the shorter lifetime and longer lifetime, respectively, and $A_i$ (i = S, L) is the amplitude of the components at t = 0. The shorter lifetime is on a time scale of several nanoseconds, and the longer lifetime is on a time scale of tens of nanoseconds. The shorter lifetime is generally attributed to the intrinsic recombination of initially populated internal core states; however, the possible origin of the longer lifetime remains relatively uncertain [51-53]. Recently, Xiao et al. reported that the longer lifetime component in the PL decay is caused by the radiative recombination of electrons and holes on the surface involving surface-localized states [53]. For QDs with high PL-QY, the amplitude $A_L$ with longer lifetime dominants the total PL. In other words, electrons and holes have an increased probability to be presented on the surfaces of QDs with high QY to contribute to this surface-related emission with a longer lifetime.

The representative PL decay curves recorded for the bare CdSe and CdS QDs and their surfaces treated by *n*-HA, *n*-BA, and OA molecules are shown in Figure 6. All of the decay curves were fitted to the biexponential equation (1). The resulting decay time constants ($\tau_i$), their fraction contribution ($Y_i$), average decay lifetimes ($\langle \tau \rangle$, calculated from (2)), the values of the goodness-of-fit parameter ($\chi^2$), and the quantum yield ($\Phi$) are listed in Table 1 for each system. The criteria for an acceptable fit have been justified in the previous article [54].

$$\langle \tau \rangle = \frac{\sum_i A_i \tau_i^2}{\sum_i A_i \tau_i} \quad (2)$$

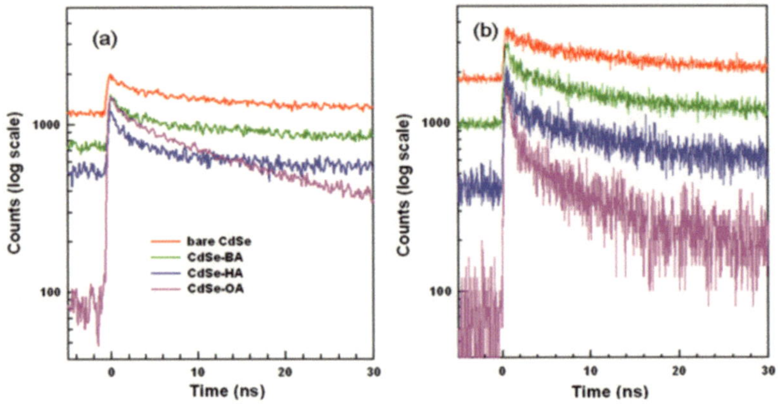


**Figure 6.** PL decay spectra for (a) bare CdSe QDs and surface modified by *n*-BA, *n*-HA and OA; (b) bare CdS QDs and surface modified by *n*-BA, *n*-HA and OA.

| Sample ID | YS (%) | τS (ns) | YL (%) | τL (ns) | <τ> (ns) | χ2 | QY (%) |
|---|---|---|---|---|---|---|---|
| CdSe | 62 | 2.93 | 38 | 16.99 | 13.90 | 0.968 | 13 |
| CdSe-BA | 44 | 2.69 | 56 | 22.87 | 21.16 | 0.931 | 44 |
| CdSe-HA | 36 | 2.50 | 64 | 28.52 | 27.30 | 1.021 | 61 |
| CdSe-OA | 89 | 0.64 | 11 | 10.76 | 7.47 | 0.926 | 6 |
| CdS | 43 | 2.27 | 57 | 22.60 | 21.18 | 0.988 | 60 |
| CdS-BA | 63 | 0.90 | 37 | 16.61 | 15.30 | 0.977 | 6.2 |
| CdS-HA | 75 | 0.88 | 25 | 18.78 | 16.61 | 0.961 | 3.2 |
| CdS-OA | 93 | 0.07 | 7 | 12.46 | 11.60 | 1.035 | 2.2 |

[a]PL decay was analyzed using biexponential model described by $I(t) = A_S e^{-t/\tau_S} + A_L e^{-t/\tau_L}$ and the fraction contribution was calculated by $Y_i = \frac{A_i}{\sum_i A_i} \times 100\%$.

**Table 1.** Fitted PL decay lifetime components.[a]

As shown in Table 1, for bare CdSe and CdS QDs, $\tau_S$ equals to 2.93 ns and 2.27 ns, and $\tau_L$ equals to 16.99 ns and 22.60 ns. The value of $\tau_S$ for the short lifetime constant is consistent with the theoretical value of approximately 3 ns, which was calculated by considering the screening of the radiating field inside the QD [55]. Moreover, for the bare CdS QD samples with 60% QY, the amplitude of $A_L$ with a longer lifetime accounts for nearly 57% of the total PL. Similar results were observed in the capping ligands for modified CdSe QDs. This larger share of the longer lifetime component is a clear indication of major surface-related emission caused by the radiative recombination of charge carriers involving surface states. Although it is difficult to identify the real origin for such radiative recombination through surface state centers in this stage, it may result from the net residual charges on atoms of CdSe and CdS QDs caused by the bonding of Cd-Se and Cd-S and the nearest neighboring atoms in the crystal lattice.

## 4. The investigation of density function theory (DFT) on capping CdS and CdSe QDs.

Here, we proposed to use the density function theory (DFT) calculations result to provide atomistic information on the adsorption energy and charge transfer of small QDs capped by different types of ligands. Theoretical modeling could provide valuable insight in atomic scale. Unfortunately, efforts in this field have been rather limited due to the high computational cost and uncertainty related to the chemical composition and morphology of the nanocrystals. Early theoretical studies of QDs simulated the surface by assuming an infinite potential barrier around the QDs. [56]. More sophisticated models of capping QDs have represented the QD core through the bulk atoms using semiempirical tight-binding [57] and pseudopotential [58] approaches, while the passivating molecules have been modeled through either single oxygen atoms or simplified model potentials [59]. Any realistic model, however, has to explicitly describe bonding between the QD and the ligands, which is lacking from the approaches mentioned above. First-principle quantum-chemical methods, such as density functional theory (DFT), are able to provide this information with a reasonable level of accuracy. Unfortunately, DFT is numerically expensive. Therefore, most DFT calculations simulate the core atoms of QDs on the basis of bulk structures, while dangling bonds on the surface are artificially terminated wit covalently bonded hydrogen atoms [60-61]. Only a few reports have focused the specifics of the adsorption energies and charge transfer of small cluster of QD interacting with ligand molecule. [46, 62]

The experimentally studied QDs retained the original crystal structure in the core. Figure 7 shows the construction of CdSe from a zinc blended lattice with bulk Cd-Se bond lengths. The analogous construction of CdSe clusters from the bulk semiconductor has been used in theoretical studies. [46, 63-65]. In addition to the uncapped ("bare") CdSe and CdS, we simulated clusters with ligands attached to the surface of the QDs. The selected capping groups, that is, *n*-butylamine (*n*-BA), *n*-hexylamine (*n*-HA), and oleic acid (OA) were used as ligands for the QDs. Thus, our simulations allowed us to study the interaction between ligand binding to Cd atoms, and the effects of these differences on the adsorption energy and optical response of the ligand capping of CdSe and CdS QDs.

In the previous studies have shown computationally that the dominant binding interaction occur between N atoms of the ligand and Cd atoms on the QD surface. We start our modeling with constructing three systems, $Cd_4Se_4$, $Cd_4Se_4$HA, $Cd_4Se_4$BA, $Cd_4Se_4$OA, $Cd_3S_5$, $Cd_3S_5$HA, $Cd_3S_5$BA, $Cd_3S_5$OA, and CdSe(111) slab (see Figure 8 (b) and (c)). The ligands were attached to the most chemically active surface atoms (all 2-coordinated Cd atoms on the cluster surface) as described the previous studies [49, 66]. Recent experiments have shown that such small "magic"-size QDs with diameter than ~2nm demonstrate great stability, the controllable size and shape, and reproducible optical properties, including an efficient blue-light emission [64]. The $Cd_{33}Se_{33}$ "magic" structure with diameter of 1.3 nm has been experimentally shown to be very stable [64], while it is the smallest cluster that supports a crystalline-like core-shell.

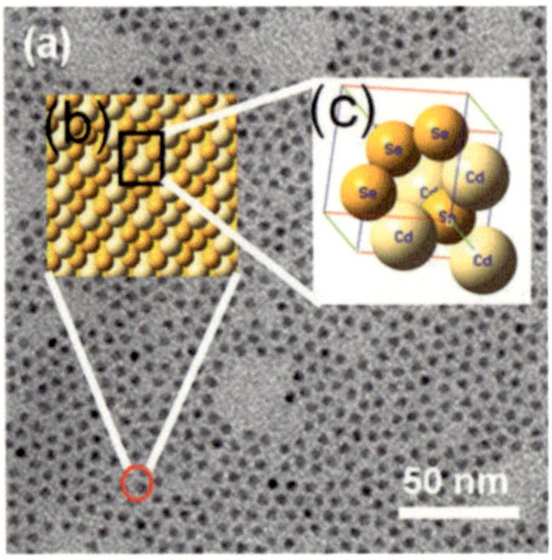


**Figure 7.** The (a) TEM image of a sample of CdSe QDs; (b) Snapshot of CdSe 5 × 5 supercell; (c) unit cell of CdSe lattice.

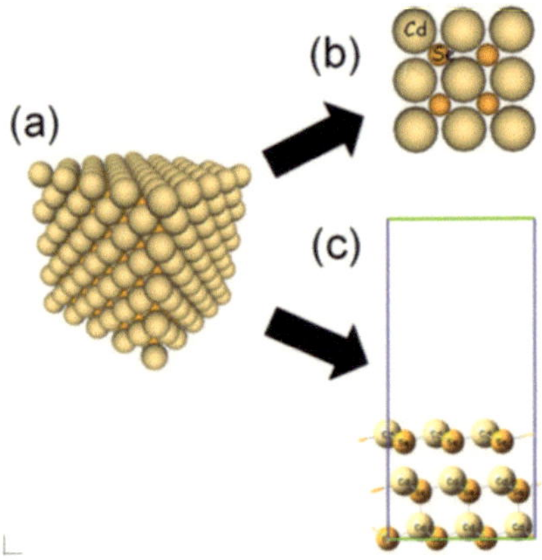


**Figure 8.** (a) The supercell of QD. (b) A finite cluster is used to model the surface (in the cluster approach) (c) A well-defined, finite vacuum space is used to model the surface (in the slab approach).

Two systems were applied in this work. To save computation resources, the $Cd_4M_4$ (M = Se and S) cluster is a reliable model for quantum-chemical studies of physical chemistry properties of CdSe and CdS QDs. The system requires reasonable computational efforts for atomistic modeling based on DFT, even when it is passivated with multiple ligands. A cluster model was used to simulate the interaction between the ligand and QD. This cluster was cut from the CdSe supercell optimized by DFT. Figure 9 shows the cluster model, in which the formula of the cluster is $Cd_4Se_4$ and the nitrogen atoms are bonded on the Cd atoms.

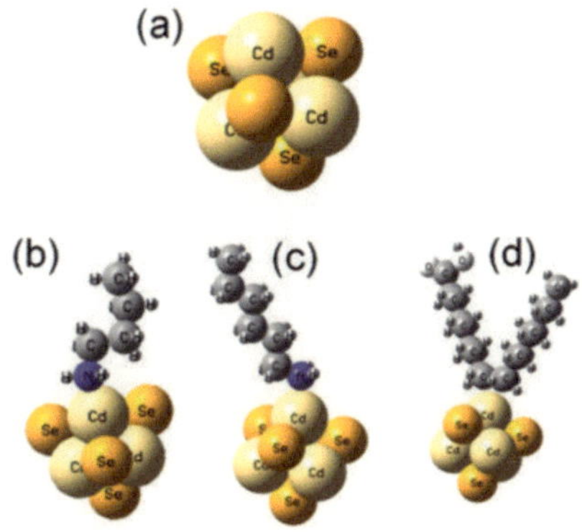


**Figure 9.** The optimized geometry of (a) CdSe cluster, (b) CdSe with *n*-BA, (c) CdSe with *n*-HA, and (d) CdSe with OA ligand.

## 4.1. CdSe

As the CdSe nanostructures were modified by the capping ligands, the capping effect of *n*-HA, *n*-BA, and OA was derived in regard to the binding energy ($E_b$) of capping molecules on a CdSe cluster and the residual surface charge on the capping molecules and the CdSe with the help of ab initio simulations. In the DFT calculation, amine and carboxylic acid were considered the main functional groups in obtaining the $E_b$ on CdSe. To simplify the simulations, we used a $Cd_4Se_4$ cluster to model the functional groups (Figure 9). During the simulations, the capping molecules were assumed to adsorb and attach on the CdSe cluster, and the binding energies of *n*-BA, *n*-HA, and OA ligands were computed using DFT simulations of an isolated cluster with a single ligand molecule. Representative snapshots for each of the capping ligands adsorbed on an isolated CdSe cluster are shown in Figure 9(b)-(d). The binding energy, $E_b$, was defined as the sum of interactions between the capping molecule and cluster atoms, and was derived as $E_b = E_{total} - E_{CdSe} - E_{capp}$, where $E_{total}$, $E_{CdSe}$ and $E_{capp}$ are the total energy of the system, cluster energy, and capping molecule energy, respectively. The negative sign of $E_b$ corresponds to the energy gain of the system because of ligand adsorption.

Table 2 lists the $E_b$ values calculated for the three capping ligands. We verified that the main contribution to the $E_b$ resulted from the interactions of the CdSe with the capping ligands through the charges on the CdSe and the ligand functional groups. The negatively charged atoms of amines adsorbed on the CdSe cluster without changing the surface structure substantially. The *n*-BA and *n*-HA adsorbed exclusively through its nitrogen atom with $E_b$ values of –0.99 and –0.93 eV for CdSe–BA and CdSe–HA, respectively. Conversely, OA yielded an $E_b$ of –0.21 eV for a CdSe cluster–OA combined with the conjugated bond (C=C) of the alkyl chain adsorbed on CdSe. Our attempts for a CdSe cluster–OA combined with carboxyl function group adsorbed on CdSe resulted in high energy and an unstable configuration with continuously varying distance between the reactive centers. This occurred because, according to the well-known hard and soft acids and bases theory [67], the $Cd^{2+}$ and $Se^{2-}$ soft ions cannot interact with hard –COOH of OA, as suggested by Chen et al. [68]. The charge analysis showed that the charge transfer between OA and CdSe was small.

| CdSe | | CdS | |
|---|---|---|---|
| ligand | $E_b$ (eV) | ligand | $E_b$ (eV) |
| (*n*-BA)$C_4NH_2$ | -0.93 | (*n*-BA)$C_4NH_2$ | -0.71 |
| (*n*-HA)$C_6NH_2$ | -0.99 | (*n*-HA)$C_6NH_2$ | -0.84 |
| OA | -0.21 | OA | -0.13 |

**Table 2.** Binding Energies ($E_b$) of different ligands on CdSe and CdS QD

Puzder et al. obtained ca. 0.91 eV from DFT calculations for trimethylamine on a $(CdSe)_{15}$ NC [46]. Similar values (0.89-1.02 eV) have been reported in the DFT study of amines at a CdSe surface [69]. Our results for organic amines are in good agreement with these values. It must be noted that the more negative the $E_b$ value the stronger is the adsorption. Indeed, in the simulation work, the more negative $E_b$ value was used as optimum to represent the adsorption strength of the functional group on a given cluster, when several other possible configurations for the adsorption on the CdSe cluster existed. Now, concerning the three capping ligands, the higher negative $E_b$ values for the amine derivatives clearly indicate a stronger adsorption of these compounds on CdSe than the OA acid ligand with a lesser negative $E_b$.

The Bader charge analyses were carried out for CdSe, *n*-BA, *n*-HA, OA, CdSe–BA, CdSe–HA and CdSe–OA to examine the variations in $E_b$ in terms of charge transfer between CdSe and capping molecules , and the charge results are listed in Table 3. In CdSe–BA and CdSe–HA systems, the charge of selenium was increased from 6.614 e (for the bare CdSe) to 6.696e and 6.700e, respectively. This is quite reasonable since the donation of charge of nitrogen atom of capping molecules to Se would easily occur. On the other hand, in CdSe-OA system, the charge of selenium atom was increased to 6.637 e lesser than that of selenium atom in CdSe-BA and CdSe-HA systems.

The charges for Cd in CdSe–BA, CdSe–HA, and CdSe–OA exhibited a decrease from 11.404e for a bare CdSe, indicating a net electron transfer from Cd atom to Se for the three capping molecules. Cd has lower electronegativity (1.7) than that of nitrogen (3.0); therefore, greater electron donation occurs to Se from the N atom of the capping molecules than that from Cd. Although the amine molecules adsorb strongly with a facile electron donation from their "–$NH_2$" functional group to Se of CdSe nanocrystals (with higher $E_b$ values), the conjugated bond (C=C) of OA forms weak bonding with CdSe and lowers the capping effect of the molecule (with lower $E_b$ values). In addition, the OA molecule with linear carbon-carbon structure can easily form a dense and stable cover layer on the surface of CdSe, thereby preventing other molecules from approaching the CdSe QDs [68]. Moreover, the structure of an OA molecule has large stereo-hindrance. All of these factors reduce the effective bonding between the OA molecule and CdSe considerably. Therefore, this study demonstrated that the improved charge transfer of amine-capped CdSe is mainly caused by the higher value of the Bader charge, implying a larger charge donation than OA, and the modification of CdSe QD by molecular capping plays a vital role in improving the CdSe charge donation.

| System | Charge(e) | Charge difference(e) | System | Charge(e) | Charge difference(e) |
|---|---|---|---|---|---|
| CdSe | Se:6.614<br>Cd: 11.404 | | CdS | S:6.723<br>Cd: 11.227 | |
| BA | N: 7.787 | | BA | N: 7.787 | |
| HA | N: 7.801 | | HA | N: 7.801 | |
| OA | O: 7.921 | | OA | O: 7.921 | |
| CdSe-BA | Se: 6.696 | Se: 6.614-6.696 = -0.082 | CdS-BA | S: 6.763 | S: 6.723-6.763 = -0.040 |
| | Cd: 11.388 | Cd: 11.404-11.388 = + 0.016 | | Cd: 11.194 | Cd: 11.227-11.194 = + 0.033 |
| | N: 7.561 | N: 7.787-7.561 = +0.226 | | N: 7.764 | N: 7.787-7.764 = +0.023 |
| CdSe-HA | Se: 6.700 | Se: 6.614-6.700 = -0.086 | CdS-HA | S: 6.775 | Se: 6.723-6.775 = -0.052 |
| | Cd: 11.375 | Cd: 11.404-11.375 = +0.029 | | Cd: 11.207 | Cd: 11.227-11.207 = +0.020 |
| | N: 7.561 | N: 7.801-7.561 = +0.24 | | N: 7.791 | N: 7.801-7.791 = +0.010 |
| CdSe-OA | Se: 6.637 | Se:6.614- 6.637 = -0.023 | CdS-OA | S: 6.727 | S: 6.723- 6.727 = -0.004 |
| | Cd: 11.267 | Cd: 11.404- 11.267 = +0.137 | | Cd: 11.184 | Cd: 11.227- 11.184 = +0.043 |
| | O: 7.906 | O: 7.921-7.906 = +0.015 | | O: 7.900 | O: 7.921-7.900 = +0.021 |

**Table 3.** Charge transfer between CdSe and CdS and ligands

### 4.2. CdS

Surface modification of CdS nanostructures by the organics was treated in regard to the adsorption of *n*-HA, *n*-BA, and OA, and the corresponding binding energy ($E_b$) of organic molecules and the residual surface charge on the organic molecules and the CdS were calculated with the help of ab initio simulations. Amine and carboxylic acid were considered the main functional groups for adsorption. A sulfur-rich structure on the surface (Cd : S=1 : 1.3) was identified through XPS analyses of the CdS QDs. To simplify the simulations, we used a $Cd_3S_5$ cluster to model its interaction with the functional groups (Figure 10(a)). During simulations, the organic molecules were assumed to adsorb and attach on the CdS cluster, and the binding energies of *n*-BA, *n*-HA, and OA ligands were computed using DFT simulations of an isolated cluster adjoined with a single ligand molecule. Representative snapshots for each of the adsorbed molecules on an isolated CdS cluster are shown in Figure 10(b)-(d). The binding energy, $E_b$, was defined as the sum of interactions between the organic molecule and cluster atoms, and was derived as $E_b = E_{total} - E_{CdS} - E_{capp}$, where $E_{total}$, $E_{CdS}$ and $E_{capp}$ are the total energy of the system, cluster energy, and organic molecule energy, respectively. The negative sign of $E_b$ corresponds to the energy gain of the system caused by ligand adsorption.

The $E_b$ values calculated for the three organic ligands are listed in Table 3. The main contribution to $E_b$ resulted from the interactions of the CdS with the organic molecules through the charges on the CdS and the ligand functional groups. Negatively charged atoms of amines adsorbed on the CdS cluster without changing the surface structure substantially. The *n*-BA and *n*-HA adsorbed exclusively through its nitrogen atom. The adsorbed amines on the CdS clusters produced binding energies of –0.71 and –0.84 eV for CdS–BA and CdS–HA, respectively. Conversely, the $E_b$ of –0.13 eV was small for the CdS–OA system with the conjugated bond (C=C) of the alkyl chain of OA adsorbed on CdS. Our attempts for a CdS cluster–OA with carboxyl function group adsorbed on CdS resulted in a high energy and unstable configuration with continuously varying distance between the reactive centers. This occurred because, according to the well-known hard and soft acids and bases theory, [67], the $Cd^{2+}$ and $S^{2-}$ soft ions cannot interact with hard –COOH of OA, as suggested by Chen et al. [68]. The more negative the $E_b$ value, the stronger the adsorption. In the simulation, the final adsorption configuration with a more negative $E_b$ value was selected as the optimal configuration, and its $E_b$ value represented the adsorption strength of the functional group on a specified cluster when several other possible configurations for the adsorption on the CdS cluster were available. Strongly adsorbed organic additives generally exhibit higher binding energies of approximately –1.0 eV. Therefore, the binding energy values obtained for the organics in this study indicate that the amines *n*-BA (–0.71 eV) and *n*-HA (–0.84 eV) adsorbed moderately, whereas the carboxylic acid OA (–0.13 eV) adsorbed weakly on CdS QDs. Weak adsorption of OA can also be expected, because the OA molecule with a linear carbon-carbon structure can easily form a dense and stable cover layer on the surface of CdS, as reported for CdSe[68], thereby preventing other molecules from approaching the CdS QDs. Moreover, the structure of an OA molecule has a large stereo-hindrance. These factors reduce the effective bonding between the OA molecule and CdS considerably.

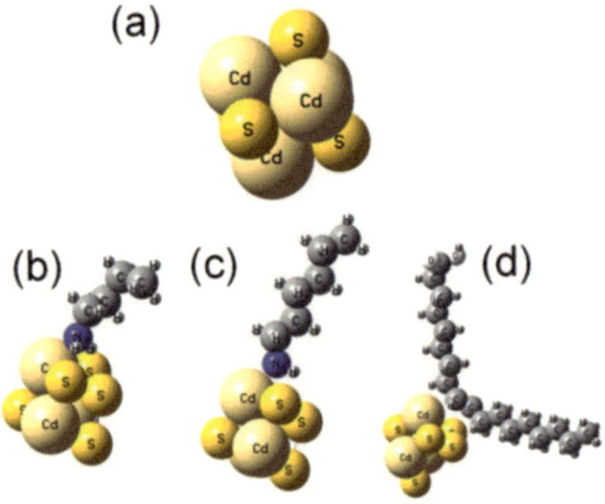


**Figure 10.** The optimized geometry of (a) CdS cluster; (b) CdS with *n*-BA; (c) CdS with *n*-HA, and (d) CdS with OA ligand.

Bader charge analyses were performed for CdS, *n*-BA, *n*-HA, OA, CdS-BA, CdS-HA, and CdS-OA to examine the variations in $E_b$ in regard to charge transfer between CdS and organic molecules; the charge results are listed in Table 2. For example, in the typical CdS-BA system, the negative charge on S increased from –6.723e (for the bare CdS) to –6.763e (net negative charge gain –0.040e) upon BA adsorption, which was caused by the donation of electrons from the N atom of the amine to S, with a loss of negative charge on N from –7.787e (for BA molecule) to –7.764e (net negative charge loss = +0.023e). Consistent with this observation, the positive charge on Cd decreased from +11.227e (for the bare CdS) to +11.194e (net positive charge loss = –0.033e), which indicates that an electron transfer occurred from the N atom of the amine to the Cd of CdS. The residual charges on S and Cd of CdS QDs were the main source of the radiative recombination surface state centers; therefore, the changes in the residual charges on S and Cd of CdS clusters, which resulted from organic molecular adsorption, can directly alter the quantity of these radiative surface states on QDs. Thus, in the case of the typical CdS-BA system, the decreased positive charge on Cd can downsize electron surface states, which reduces the surface-related emission, the longer lifetime component, and eventually, the overall PL-QY. Conversely, the net gain of negative charge on S, which can multiply hole surface states, predicts higher surface-related emission with an increased longer lifetime component and higher overall PL-QY, which is in contrast to the experimental observations, indicating that hole surface states-related phenomena do not occur. This occurred because, although abundant quantities of photoexcited holes were available on the organics-modified QDs surface, fewer electron surface states and photoexcited electrons necessary for radiative electron-hole recombination were available on the CdS QDs surface because of the decreased Cd atom positive charge, thereby preventing such an increase of PL-QY.

The proposed mechanism can be extended to explain the behavior of the other amine, *n*-HA, and the acid OA towards decreasing the longer lifetime component of photoexcited charges and the PL-QY of CdS QDs. Notably, the acid OA caused more changes compared to the amines, despite the fact that OA had low binding energy (–0.13 eV, Table 2), and weakly adsorbed on the CdS in comparison to the amines. Table 3 shows that the positive charge loss on Cd was the highest (–0.043e) by the OA molecule, compared to BA (–0.033e) and HA (–0.020e). Consequently, more electron surface states can be reduced on the surface by OA, and this trend of decreasing electron surface states is consistent with the decreasing longer lifetime component and lower PL-QY by this ligand. This confirms that the net charge transfer between the organic molecule and the CdS QDs is the crucial factor, rather than the binding energy of the molecule, in effecting the carrier recombination dynamics by controlling the radiative recombination surface state centers.

## 5. Conclusion

Based on the TRPL analysis and DFT computation, this study demonstrated that the amine can be used as the capping ligands for CdSe QDs to enhance PL-QYs, whereas the amine used as the capping ligands for CdS QDs can be used to PL-QYs. The PL-QYs of CdSe and CdS QDs decreased considerably when using oleic acid as capping ligands. We propose that the interactions between capping ligands and CdSe or CdS QDs may be attributed to $NH_2$ group for amine molecules and –COOH group for fatty acid molecules. An improved understanding of this interaction will facilitate the design of superior conjugates for CdSe and CdS QDs for various applications.

## Author details

Liang-Yih Chena* and Chia-Hung Tseng
*Department of Chemical Engineering, National Taiwan University of Science and Technology, Taipei, Taiwan*

Hung-Lung Chou
*Graduate Institute of Applied Science and Technology, National Taiwan University of Science and Technology, Taipei, Taiwan*

Ching-Hsiang Chen
*Protrustech Corporation Limited, Tainan, Taiwan*

## Acknowledgement

We thank the NCHC HPC and NTUST for providing massive computing time. Financially support from the National Science Council under Contract No. NSC 99-2811-M-011-005 is gratefully acknowledged.

## 6. References

[1] G. Markovich, C. P. Collier, S. E. Henrichs, F. Remacle, R. D. Levine, and J. R. Heath (1999) Architectonic Quantum Dot Solids. Acc. Chem. Res. 32: 415-423.

[2] A. P. Alivisatos (1996) Semiconductor Clusters, Nanocrystals, and Quantum Dots Science. 271: 933-937.

[3] M. Bruchez, M. M. Jr., P. Gin, S. Weiss, and A. P. Alivisatos (1998) Semiconductor Nanocrystals as Fluorescent Biological Labels Science. 281: 2013-2016.

[4] W. C. W. Chan and S. Nie (1998) Quantum Dot Bioconjugates for Ultrasensitive Nonisotopic Detection Science. 281: 2016-2018.

[5] J. Hranisavljevic, N. M. Dimitrijevic, G. A. Wurtz, and G. P. Wiederrecht (2002) Photoinduced Charge Separation Reactions of J-Aggregates Coated on Silver Nanoparticles. J. Am. Chem. Soc. 124: 4536-4537.

[6] J. K. Jaiswal, H. Mattoussi, J. M. Mauro, and S. M. Simon (2003) Long-term Multiple Color Imaging of Live Cells Using Quantum Dot Bioconjugates Nat. Biotechnol. 21: 47-51.

[7] S. Kim, Y. T. Lim, E. G. Soltesz, A. M. D. Grand, J. Lee, A. Nakayama, J. A. Parker, T. Mihaljevic, R. G. Laurence, D. M. Dor, L. H. Cohn, M. G. Bawendi, and J. V. Frangioni (2004) Near-Infrared Fluorescent Type II Quantum Dots for Sentinel Lymph Node mapping Nat. Biotechnol. 22: 93-97.

[8] X. Gao, Y. Cui, R. M. Levenson, L. W. K. Chung, and S. Nie (2004) In Vivo Cancer Targeting and Imaging with Semiconductor Quantum Dots Nat. Biotechnol. 22: 969-976.

* Corresponding Author

[9] W. U. Huynh, X. Peng, and A. P. Alivisatos (1999) CdSe Nanocrystal Rods/Poly(3-hexylthiophene) Composite Photovoltaic Devices. Adv. Mater. 11: 923-927.
[10] W. U. Huynh, J. J. Dittmer, and A. P. Alivisatos (2002) Hybrid Nanorod-Polymer Solar Cells Science. 295: 2425-2427.
[11] R. D. Schaller and V. I. Klimov (2004) High Efficiency Carrier Multiplication in PbSe Nanocrystals: Implications for Solar Energy Conversion. Phys. Rev. Lett. 92: 186601-1~186601-4.
[12] J. Liu, T. Tanaka, K. Sivula, A. P. Alivisatos, and J. M. J. Fréchet (2004) Employing End-Functional Polythiophene To Control the Morphology of Nanocrystal–Polymer Composites in Hybrid Solar Cells. J. Am. Chem. Soc. 126: 6550-6551.
[13] R. J. Ellingson, M. C. Beard, J. C. Johnson, P. Yu, O. I. Micic, A. J. Nozik, A. Shabaev, and A. L. Efros (2005) Highly Efficient Multiple Exciton Generation in Colloidal PbSe and PbS Quantum Dots. Nano Lett. 5: 865-871.
[14] W. Cai, D.-W. Shin, K. Chen, O. Gheysens, Q. Cao, S. X. Wang, S. S. Gambhir, and X. Chen (2006) Peptide-Labeled Near-Infrared Quantum Dots for Imaging Tumor Vasculature in Living Subjects. Nano Lett. 6: 669-676.
[15] N. J. Smith, K. J. Emmett, and S. J. Rosenthal (2008) Photovoltaic Cells Fabricated by Electrophoretic Deposition of CdSe Nanocrystals Appl. Phys. Lett. 93: 043504.
[16] B. O. Dabbousi, M. G. Bawendi, O. Onitsuka, and M. F. Rubner (1995) Electroluminescence from CdSe Quantum-Dot/Polymer Composites. Appl. Phys. Lett. 66: 1316.
[17] M. C. Schlamp, X. Peng, and A. P. Alivisatos (1997) Improved Efficiencies in Light Emitting Diodes made with CdSe(CdS) Core/Shell type Nanocrystals and a Semiconducting Polymer. J. Appl. Phys. 82: 5837.
[18] H. Mattoussi, L. H. Radzilowski, B. O. Dabbousi, E. L. Thomas, M. G. Bawendi, and M. F. Rubner (1998) Electroluminescence from Heterostructures of Poly(phenylene vinylene) and Inorganic CdSe Nanocrystals J. Appl. Phys. 83: 7965.
[19] S. Coe, W.-K. Woo, M. Bawendi, and V. Bulovi (2002) Electroluminescence from Single Monolayers of Nanocrystals in Molecular Organic Devices Nature. 420: 800-803.
[20] R. A. M. Hikmet, P. T. K. Chin, D. V. Talapin, and H. Weller (2005) Polarized-Light-Emitting Quantum-Rod Diodes. Adv. Mater. 17: 1436-1439.
[21] A. H. Mueller, M. A. Petruska, M. Achermann, D. J. Werder, E. A. Akhadov, D. D. Koleske, M. A. Hoffbauer, and V. I. Klimov (2005) Multicolor Light-Emitting Diodes Based on Semiconductor Nanocrystals Encapsulated in GaN Charge Injection Layers. Nano Lett. 5: 1039-1044.
[22] J. Zhao, J. A. Bardecker, A. M. Munro, M. S. Liu, Y. Niu, I. K. Ding, J. Luo, B. Chen, A. K. Y. Jen, and D. S. Ginger (2006) Efficient CdSe/CdS Quantum Dot Light-Emitting Diodes Using a Thermally Polymerized Hole Transport Layer. Nano Lett. 6: 463-467.
[23] P. O. Anikeeva, C. F. Madigan, J. E. Halpert, M. G. Bawendi, and V. Bulović (2008) Electronic and Excitonic Processes in Light-Emitting Devices Based on Organic Materials and Colloidal Quantum Dots. Phys. Rev. B. 78: 085434.
[24] A. P. Alivisatos (1996) Perspectives on the Physical Chemistry of Semiconductor Nanocrystals. J. Phys. Chem. 100: 13226-13239.

[25] M. Marandi, N. Taghavinia, A. Iraji, and S. M. Mahdavi (2005) A Photochemical Method for Controlling the Size of CdS Nanoparticles. Nanotechnology. 16: 334-338.
[26] W. W. Yu and X. Peng (2002) Formation of High-Quality CdS and Other II-VI Semiconductor Nanocrystals in Noncoordinating Solvents: Tunable Reactivity of Monomers. Angew Chem. Int. Ed. 41: 2368-2371.
[27] D. Pan, S. Jiang, L. An, and B. Jiang (2004) Controllable Synthesis of Highly Luminescent and Monodisperse CdS Nanocrystals by a two-Phase Approach under Mild Conditions. Adv. Mater. 16: 982-985.
[28] J. He, J. Mi, H. Li, and W. Ji (2005) Observation of Interband Two-Photon Absorption Saturation in CdS Nanocrystals. J. Phys. Chem. B. 109: 19184-19187.
[29] J. Ouyang, J. Kuijper, S. Brot, D. Kingston, X. Wu, D. M. Leek, M. Z. Hu, J. A. Ripmeester, and K. Yu (2009) Photoluminescent Colloidal CdS Nanocrystals with High Quality via Noninjection One-Pot Synthesis in 1-Octadecene. J. Phys. Chem. C. 113: 7579-7593.
[30] C. B. Murray, D. J. Norris, and M. G. Bawendi (1993) Synthesis and characterization of nearly monodisperse CdE (E = sulfur, selenium, tellurium) semiconductor nanocrystallites. J. Am. Chem. Soc. 115: 8706-8715.
[31] M. E. Wankhede and S. K. Haram (2003) Synthesis and Characterization of Cd–DMSO Complex Capped CdS Nanoparticles. Chem. Mater. 15: 1296-1301.
[32] E. Jang, S. Jun, Y. Chung, and L. Pu (2004) Surface Treatment to Enhance the Quantum Efficiency of Semiconductor Nanocrystals. J. Phys. Chem. B. 108: 4597-4600.
[33] P. Thangadurai, S. Balaji, and P. T. Manoharan (2008) Surface Modification of CdS Quantum Dots Using Thiols—Structural and Photophysical Studies Nanotechnology. 19: 435708-1~435708-8.
[34] J. Aldana, Y. A. Wang, and X. Peng (2001) Photochemical Instability of CdSe Nanocrystals Coated by Hydrophilic Thiols. J. Am. Chem. Soc. 123: 8844-8850.
[35] D. V. Talapin, A. L. Rogach, A. Kornowski, M. Haase, and H. Weller (2001) Highly Luminescent Monodisperse CdSe and CdSe/ZnS Nanocrystals Synthesized in a Hexadecylamine–Trioctylphosphine Oxide–Trioctylphospine Mixture. Nano Lett. 1: 207-211.
[36] V. Swayambunathan, D. Hayes, K. H. Schmidt, Y. X. Liao, and D. Meisel (1990) Thiol surface complexation on growing cadmium sulfide clusters. J. Am. Chem. Soc. 112: 3831-3837.
[37] T. Uchihara, S. Maedomari, T. Komesu, and K. Tanaka Influences of Proton-Dissociation Equilibrium of Capping Agents on the Photo-Chemical Events of the Colloidal Solutions Containing the Thiol-Capped Cadmium Sulfide Particles J. Photochem. and Photobio. 161: 227-232.
[38] X. Peng, T. E. Wilson, A. P. Alivisatos, and P. G. Schultz (1997) Synthesis and Isolatin of a Homodimer of Cadmium Selenide Nanocrystals. Angew Chem. Int. Ed. 36: 145-147.
[39] S. Pathak, S.-K. Choi, N. Arnheim, and M. E. Thompson (2001) Hydroxylated Quantum Dots as Luminescent Probes for in Situ Hybridization. J. Am. Chem. Soc. 123: 4103-4104.
[40] G. Kalyuzhny and R. W. Murray (2005) Ligand Effects on Optical Properties of CdSe Nanocrystals. J. Phys. Chem. B. 109: 7012-7021.

[41] C. Bullen and P. Mulvaney (2006) The Effects of Chemisorption on the Luminescence of CdSe Quantum Dots. Langmuir. 22: 3007-3013.

[42] A. M. Munro, I. J.-L. Plante, M. S. Ng, and D. S. Ginger (2007) Quantitative Study of the Effects of Surface Ligand Concentration on CdSe Nanocrystal Photoluminescence. J. Phys. Chem. C. 111: 6220-6227.

[43] S. A. Gallagher, M. P. Moloney, M. Wojdyla, S. J. Quinn, J. M. Kelly, and Y. K. Gun'ko (2010) Synthesis and Spectroscopic Studies of Chiral CdSe Quantum Dots J. Mater. Chem. 20: 8350-8355.

[44] O. Chen, Y. Yang, T. Wang, H. Wu, C. Niu, J. Yang, and Y. C. Cao (2011) Surface-Functionalization-Dependent Optical Properties of II–VI Semiconductor Nanocrystals. J. Am. Chem. Soc. 133: 17504-17512.

[45] G. Kalyuzhny and R. W. Murray. (2005) Ligand Effects on Optical Properties of CdSe Nanocrystals. J. Phys. Chem. B. 109: 7012-7021.

[46] A. Hoshino, K. Fujioka, T. Oku, M. Suga, Y. F. Sasaki, T. Ohta, M. Yasuhara, K. Suzuki, and K. Yamamoto (2004) Physicochemical Properties and Cellular Toxicity of Nanocrystal Quantum Dots Depend on Their Surface Modification. Nano Lett. 4: 2163-2169.

[47] T. Tsuruoka, K. Akamatsu, and H. Nawafune (2004) Synthesis, Surface Modification, and Multilayer Construction of Mixed-Monolayer-Protected CdS Nanoparticles. Langmuir. 20: 11169-11174.

[48] F. Dubois, B. Mahier, B. Dubertret, E. Doris, and C. Mioskowski (2006) A Versatile Strategy for Quantum Dot Ligand Exchange. J. Am. Chem. Soc. 129: 482-483.

[49] H.-L. Chou, C.-H. Tseng, K. C. Pillai, B.-J. Hwang, and L.-Y. Chen (2010) Adsorption and Binding of Capping Molecules for Highly Luminescent CdSe Nanocrystals - DFT Simulation Studies. Nanoscale. 2: 2679-2684.

[50] M. Jones and G. D. Scholes (2010) On the use of Time-Resolved Photoluminescence as a Probe of Nanocrystal Photoexcitation Dynamics J. Mater. Chem. 20: 3533.

[51] G. Schlegel, J. Bohnenberger, I. Potapova, and A. Mews (2002) Fluorescence Decay Time of Single Semiconductor Nanocrystals. Phys. Rev. Lett. 88: 137401.

[52] M. Nirmal, D. J. Norris, M. Kuno, M. G. Bawendi, A. L. Efros, and M. Rosen (1995) Observation of the "Dark Exciton" in CdSe Quantum Dots. Phys. Rev. Lett. 75: 3728-3731.

[53] X. Wang, L. Qu, J. Zhang, X. Peng, and M. Xiao (2003) Surface-Related Emission in Highly Luminescent CdSe Quantum Dots. Nano Lett. 3: 1103-1106.

[54] M. Jones, S. Kumar, S. S. Lo, and G. D. Scholes (2008) Exciton Trapping and Recombination in Type II CdSe/CdTe Nanorod Heterostructures. J. Phys. Chem. C. 112: 5423-5431.

[55] B. L. Wehrenberg, C. Wang, and P. Guyot-Sionnest (2002) Interband and Intraband Optical Studies of PbSe Colloidal Quantum Dots. J. Phys. Chem. B. 106: 10634-10640.

[56] A. I. Ekimov, F. Hache, M. C. Schanne-Klein, D. D. Ricard, C. Flytzanis, I. A. Kudryavtsev, T. V. Yazeva, A. V. Rodina, and A. L. Efros (1993) Absorption and Intensity-Dependent Photoluminescence Measurements on CdSe Quantum Dots: Assignment of the First Electronic Transitions. J. Opt. Soc. Am. B. 10: 100-107.

[57] S. Pokrant and K. B. Whaley (1999) Tight-Binding Studies of Surface Effects on Electronic Structure of CdSe Nanocrystals: the Role of Organic Ligands, Surface Reconstruction, and Inorganic Capping Shells. Eur. Phys. J. D. 6: 255-267.

[58] L.-W. Wang and A. Zunger (1996) Pseudopotential Calculations of Nanoscale CdSe Quantum Dots. Phys. Rev. B. 53: 9579-9582.

[59] F. A. Reboredo and A. Zunger (2001) Surface-Passivation-Induced Optical Changes in Ge Quantum Dots. Phys. Rev. B. 63: 235314.

[60] G. M. Dalpian, M. L. Tiago, M. L. d. Puerto, and J. R. Chelikowsky (2006) Symmetry Considerations in CdSe Nanocrystals. Nano Lett. 6: 501-504.

[61] M. L. del Puerto, M. L. Tiago, and J. R. Chelikowsky (2006) Excitonic Effects and Optical Properties of Passivated CdSe Clusters. Phys. Rev. Lett. 97: 096401.

[62] J. Y. Rempel, B. L. Trout, M. G. Bawendi, and K. F. Jensen (2006) Density Functional Theory Study of Ligand Binding on CdSe (0001), $(000\bar{1})$, and $(11\bar{2}0)$ Single Crystal Relaxed and Reconstructed Surfaces: Implications for Nanocrystalline Growth. J. Phys. Chem. B. 110: 18007-18016.

[63] A. Puzder, A. J. Williamson, F. Gygi, and G. Galli (2004) Self-Healing of CdSe Nanocrystals: First-Principles Calculations. Phys. Rev. Lett. 92: 217401.

[64] S. Kilina, S. Ivanov, and S. Tretiak (2009) Effect of Surface Ligands on Optical and Electronic Spectra of Semiconductor Nanoclusters. J. Am. Chem. Soc. 131: 7717-7726.

[65] H. Kamisaka, S. V. Kilina, K. Yamashita, and O. V. Prezhdo (2008) Ab Initio Study of Temperature and Pressure Dependence of Energy and Phonon-Induced Dephasing of Electronic Excitations in CdSe and PbSe Quantum Dots†. J. Phys. Chem. C. 112: 7800-7808.

[66] H.-L. Chou, C.-H. Tseng, K. C. Pillai, B.-J. Hwang, and L.-Y. Chen (2011) Surface Related Emission in CdS Quantum Dots. DFT Simulation Studies. J. Phys. Chem. C. 115: 20856-20863.

[67] R. G. Pearson (1963) Hard and Soft Acids and Bases. J. Am. Chem. Soc. 85: 3533-3539.

[68] J. Chen, J. L. Song, X. W. Sun, W. Q. Deng, C. Y. Jiang, W. Lei, J. H. Huang, and R. S. Liu (2009) An Oleic Acid-Capped CdSe Quantum-Dot Sensitized Solar Cell Appl. Phys. Lett. 94: 153115.

[69] P. Schapotschnikow, B. Hommersom, and T. J. H. Vlugt (2009) Adsorption and Binding of Ligands to CdSe Nanocrystals. J. Phys. Chem. C. 113: 12690-12698.

Chapter 9

# The Synthesis of Nano-Crystalline Metal Oxides by Solution Method

Xuejun Zhang and Fuxing Gan

Additional information is available at the end of the chapter

http://dx.doi.org/10.5772/45643

## 1. Introduction

The performance of materials in many of their uses in industries and scientific researches is directly dependent on their crystal structure, which correlates to the chemical and physical properties of the materials. And also, as the size of crystal particles decrease to nanometer scale, nanocrystals exhibit some unexpected properties that are evidently different in physics and chemistry from their bulk crystals and their cluster compounds as well.

One of the most important applications of nano-materials is as catalyst to be widely used in petroleum and chemical industries, which has been a hot research area attracting high attention from researchers around the world. A lot of nanomaterials that have shown highly catalytic activity are nano-sized metal oxide crystals or doped metal oxide crystals. As well known, the nature of semiconductor is one of major features for solid catalyst, especially for solid photo-catalyst. The properties could exhibit or work as catalysts only when the metal oxide bears crystal structure. Therefore, the crystallization of oxide or doped oxide is a key step in the preparation of catalyst.

Nowadays the current processes of preparing metal oxide nanocrystal are mainly involved in sol-gel method and some modified sol-gel methods. The products synthesized by the methods, however, are metal hydroxides, which have to undergo a firing treatment (at over 350 °C) in order to have them crystallized and to be endowed with semiconductive and catalytic properties. But, the formation of oxide crystals in the roasting process involves a phase transition process, in which a new grain boundary forms and expands at high temperature, leading to size increase of the particle obtained in solution synthesis or even to a new matrix element phase from which the doping element is excluded. In addition, the process of phase transformation in calcination is unfavourable for the preparation of nuclear-shell structure of nano materials such as a magnetic nuclear coated with $TiO_2$, $SiO_2$ or $SnO_2$, resulting in tow-phase separation and a failure of coating on magnetic nuclear. For the synthesis of nano-crystalline metal oxides with alterable valence, the calcination in the air causes the valence of metal to rise by oxidation and the original crystalline structure to change. And also, on the surface of the directly synthesized metal oxide nanocrystals without high-temperature burning there exists a large number of hydroxyl groups, which are more conducive to water molecules, organic solvents, or organic compounds compatible and to surface modification and functionalization of nanocrystals. See Section 4.4.

A modified sol-gel method, "precipitation–condensation with non-aqueous ion exchange (P-CNAIE)", associating with a drying method, "azeotropic drying of iso-amyl acetate (AD-IAA)", was put forward in 2005 (Zhang 2005), that is, in ethanol a strongly basic anion-resin was used as an exchanger to remove by-product $Cl^-$ and as a reactant to provide $OH^-$ for hydrolysis. The high-purity metal hydroxide tends to dehydrate in an intermolecular manner with the assistance of super water-absorbable ethanol to form crystal.

## 2. Method of synthesis

The solution chemistry method, usually referred to sol-gel method, is a significant process to synthesize the precursors of many nanoscale metal oxides. The method is widely used for it can achieve uniform doping of multi-elements no matter whether at atomic, molecular or nanometer levels at the gelatination phase. Generally, sol–gel processes have associated problems, such as difficulty in removing chlorine, and in accurate and repeated doping.

The new solution chemistry method proposed by us involves a precipitation–condensation process, with non-aqueous ion exchange in ethanol used for the removal of chlorine and for providing hydroxyl ion. (Zhang, 2005; Zhang et al., 2006, 2008; Yang et al. 2007).

### 2.1. The underlying principle of the hydrolysis method

The method involving P-CNAIE for removal of chlorine is based on a reaction that occurs in low-polar solvents containing limited water under a slightly alkalinity condition. Generally, as the soluble metal salts dissolved in water the solution become acidic and metal ions are hydrolyzed. Such tendency is much stronger for high valence metal ions because the acidity of high valence metal ions is stronger than that of low valence metal ions. This means the $K_{sp}$ of hydroxides of high valence metal ions are much smaller than that of low valence metal ions. Different kind of metal ion or the metal ions with different valence state have different $K_{sp,\ M(OH)n}$, which results in a part precipitation of the metal ion with low $K_{sp,\ M(OH)n}$ in aqueous solution. That a limited amount of water is added in the organic solvent will control the hydrolysis degree of metal ions, and in this case, addition of a limited amount of ammonia water instead can both control the hydrolysis degree and avoid part precipitation of a metal ion as a doped nano- material is synthesized.

Here $MCl_4$ is referred to as matrix metal chloride and $NCl_3$ as metal dopant. In the synthesis, the hydrolysis takes place limitedly under the assistance of ammonia water in absolute alcohol. The acidification is attributed to the HCl coming from $MCl_4$ and $NCl_3$, being caused by the first-order hydrolysis of $MCl_4$ and $NCl_3$.

$$MCl_4 + H_2O \rightarrow M(OH)Cl_3 + HCl \quad (1)$$

$$NCl_3 + H_2O \rightarrow N(OH)Cl_2 + HCl \quad (2)$$

The product of the second-order hydrolysis of $MCl_4$ and $NCl_3$ is generally a white suspended precipitate. In an organic solvent, the polar product is precipitated much more readily, and the size of the particles is smaller because of the solubility of polar compound in organic solvent is much lower than that in water. (Wuhan U. 2000) The limited amount of ammonia water controls the rate of hydrolysis and the pH of the solution to slightly alkaline, preventing a part precipitation of either $MCl_4$ or $NCl_3$:

$$MCl_4 + 2H_2O \rightarrow M(OH)_2Cl_2 \downarrow + 2HCl \quad (3)$$

$$NCl_3 + 2H_2O \rightarrow N(OH)_2Cl \downarrow + 2HCl \quad (4)$$

The third- and fourth-order hydrolysis of $MCl_4$ and $NCl_3$ is very weak. From the third-order hydrolysis on, the pH of the solution is almost maintained at around 7 (if pH > 9, the products might be $(NH_4)_2MO_3$ and $(NH_4)NO_2$, instead of $M(OH)_4$ and $N(OH)_3$, for some amphoteric elements). The third- and fourth-order hydrolysis is reversible.

$$M(OH)_2Cl_2 \downarrow + 2H_2O \rightleftarrows M(OH)_4 \downarrow + 2HCl \quad (5)$$

$$N(OH)_2Cl \downarrow + H_2O \rightleftarrows N(OH)_3 \downarrow + HCl \quad (6)$$

To maintain reactions (5) and (6) and prevent the formation of $(NH_4)_2MO_3$ and $(NH_4)NO_2$, chlorine has to be removed under neutral or slightly alkaline conditions. This is often performed by repeatedly using a fresh neutral solvent to wash the precipitate. This process consumes considerable amount of solvent and time. However, in the process of hydrolysis using anion-exchange resin, because the affinity of the anion-exchange resin for $Cl^-$ is over 25 times (Luliang et al., 2000) that for $OH^-$, the resin readily exchanges $Cl^-$ and supplies $OH^-$ for the hydrolysis.

$$Resin-OH^- + Cl^- \xrightarrow{NH_3} Resin-Cl^- + OH^- \quad (7)$$

This speeds up the hydrolysis and shortens the duration of time for formation of hydroxides of M(IV) and N(III).

The reaction for hydrolysis associates with anion resin can be written as

$$MCl_4 + 4Resin-OH^- \xrightarrow{NH_3,\,H_2O} 4Resin-Cl^- + M(OH)_4(Clusters)\downarrow \quad (8)$$

$$NCl_3 + 3Resin-OH^- \xrightarrow{NH_3,\,H_2O} 3Resin-Cl^- + N(OH)_3(Clusters)\downarrow \quad (9)$$

Instead of complete dispersion, the hydroxides of M(IV) and N(III) showed a tendency to dehydrate, and formed condensates. (D'Souza et al., 2000) Starting from the second-order hydrolysis, when the monomer hydroxide has two condensable functional groups, $f = 2$, more and more linear condensates are formed. From the third-order hydrolysis, the average functionality of the monomer becomes more than two, $\overline{f} > 2$. This allows cross-linking between linear condensates, and also gelation.

## 2.2. The operation of the method

In an airtight flask containing 200 mL anion-exchange resin (say the DOWEX Monosphere 550A UPW(OH), Dow Chemical Company, Midland, MI), 100-200 mL alcohol, and 10–15 mL of ammonia water, 200-300 mL ethanol solution containing $MCl_4$ (15-25%, w/v), or/and $NCl_3$ (desired%, w/v) are added dropwise with fast stirring. $NH_3$ gas was aerated in the reaction solution to catalyze the hydrolysis. The reaction apparatus is shown as Fig. 1.

The reaction solution is held close to neutral, pH 6-8, by adjusting the speed of addition. After the addition is complete, a solution containing a white suspended precipitate is separated with resin through a glass-sand funnel or a 120 mesh stainless steel screen. The filtrate reacts repeatedly with 50-100 mL fresh anion-exchange resin on a shaker, to continue removal of chlorine and promote further hydrolysis for five or six times until the upper solution does not become clear upon standing and $Cl^-$ in solution is not detected by More Essay. The final chlorine-free (checked by $AgNO_3$ solution) colloid solution is held idle on a bench for ca. 48–72 h and separated into an upper lightly turbid solution and a lower dense precipitate. The upper, lightly turbid solution is then removed and kept aside for final recovery of all precipitate. Iso-amyl acetate (70–100 mL) is added to the lower dense precipitate solution to make a co-boiling system. A dispersive fine powder is obtained by co-distilling off water absorbed on the colloid and solvents, or azeotropic drying.

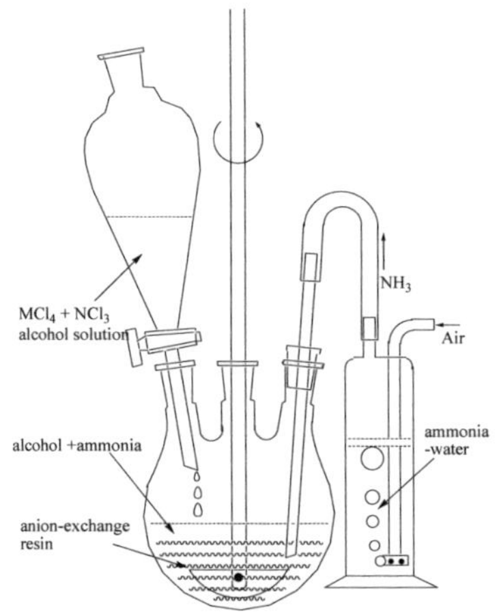


**Figure 1.** The glass apparatus for preparation of doped nanocrystal by hydrolysis of $MCl_4$ and/or $NCl_3$.

All the exchanged ion-exchange resins are combined and repeatedly washed with fresh solvent to collect any residual precipitate on the surface of the resin. The washed solvent is applied to a short column of ion exchange to remove any remaining chlorine, and is then combined with the upper lightly turbid solution. The resulting dried powder is then dispersed in the combined solution on a shaker, and a chlorine-free solution containing powder is distilled to separate the solvent and leave behind the hydrolysis product, as a high-dispersively fine powder.

## 2.3. The drying of nano-crystalline metal oxides

To coupling with the method "P-CNAIE", a drying method of "azeotropic drying of iso-amyl acetate (AD-IAA)" was presented in 2005 (Zhang, 2005; Yang et al., 2007)

Although the solution chemistry method is a significant process to synthesize many nanoscale metal oxides, the drying methods plays a key role in the successful preparation of nanoparticles. If the surface tension is not reduced (Sun & Berg, 2002) during the drying process of precipitate colloids, the colloid particles will aggregate to form rigid gel pieces in the end. It has been known that gels can be formed either through the condensation of polymers or through the aggregation of particles (Diao et al., 2002). According to cluster–cluster growth models, clusters stick together randomly with a certain probability upon colliding (Brinker & Scherer, 1990). It is found that cluster–cluster growth resulting from colliding may occur easily at the drying stage. Although dried gel bulks can be ground to powder, the mechanical force cannot crush them into the powder as fine as the particles synthesized. In addition, the ground powder displays structural features different from the original colloid-precipitated particles. Thus, along with the synthesis process, the drying process is critical in determining the dispersivity and size of the final dried products in the solution chemistry process (Richards & Khaleel, 2001) .

Several drying methods, such as supercritical drying (Boujday et al., 2004; Park et al., 2002), freeze drying (Vidal et al., 2005; Shlyakhtin et al., 2000), microwave drying (Hwang et al., 1997), and azeotropic distillation (Hu et al., 1998; Frazee & Harris, 2001; Luan et al., 1998) have been developed to remove the trace of water adsorbed on the surface of colloid particles in order to prevent or reduce aggregation caused by shrinkage of water films between precipitated particles. The water film is formed by adsorbing water on the surface as colloid particles contact. See Scheme 1. Supercritical, freeze, and microwave drying methods reduce agglomerates by eliminating or reducing the surface tension of the water films between colloids. Azeotropic distillation, however, removes the adsorbed water on the surface as colloids disperse in an azeotropic solution, thereby preventing the formation of water films between colloids and aggregation of particles. Some surfactants can prevent colloids from aggregating, but they have a high boiling point and are adsorbed on the dried powder so tightly that it is hard to be removed by the normal drying process.

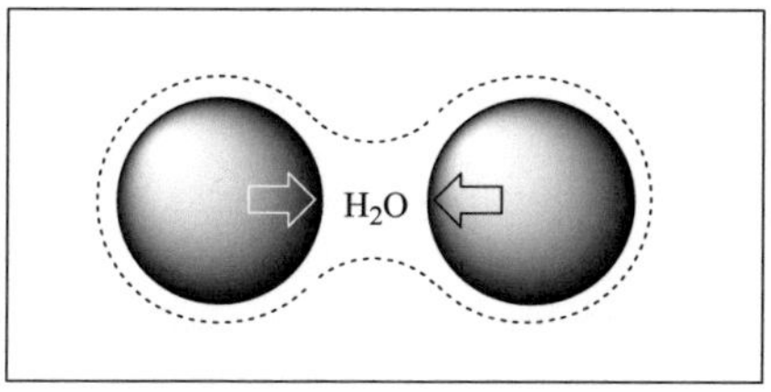


**Scheme 1.** Particles come close up as water films between particles shrink at final drying stage.

### *2.3.1. The effects of boiling point and molecular structures of azeotropic agent on the dispersivity and fluffiness of the dried powder*

We have ever presented our results from a series of azeotropic solvents such as alcohols, ethers, and esters. It is found that the molecular structures of azeotropic agents have a strong influence on the dispersivity and fluffiness of the dried products.

Experiments have demonstrated the existence of water on the surface of colloid particles by the separated water phase below the organic solvent layer in the bottle collecting condensed fluid at the azeotropic distillation stage. The water is introduced into the hydrolyzed solution by the added ammonia water, and from the condensation reaction of hydrolysate and the crystal water of metal chlorides. The surface-hydrated layer of particle confers the polymer precipitate colloid with a stable dispersivity in the ethanol. During evaporation, the ethanol is evaporated first and water molecules are left on the surface of the precipitate particles. As colloid particles come together, surface-hydrated waters combine with each other to form a water film between the particles. The film water, like capillary one, has capillary forces that can cause water film to shrink and particles eventually to contact each other and form hard brown agglomerates, for Sb-doped $SnO_2$ nano particles see Fig. 2(A).

**Figure 2.** Images of antimony-doped stannic nanocrystal that is dried directly by distilling ethanol off (A) or dried by azeotropic distillation with 1,4-butanediol (B); ethylene glycol–monomethyl ether (EGMMeE) (C); n-butanol (D); pentanol (E); iso-amyl alcohol (F); tetrachloroethylene (TCE) (G); n-butyl ether (H); and n-butyl acetate (I). The scale of bar in pictures is 1.0 cm. (Zhang, 2005; Yang et al., 2007)

To remove the water molecules on the particles and keep the colloid particle dispersed, some organic solvents in Table 1 are selected. The characterization of the dried samples in Fig. 2 is summarized in Table 1. The number of symbol + stands for different grade color, agglomerates, and hardness. It can be seen that when short-chain organic solvents with two oxygen groups are used as azeotropic agents, the properties of the dried products (Fig. 2(B) and (C)) are similar to that without azeotropic treatment (Fig. 2(A)), meaning that they play the same roles as water molecule, leading to semi-transparent xerogel pieces (Figs. 2(A) and (B)). Here, the rigid gel is formed through the aggregation (Brinker & Scherer, 1990).

Even though the dense precipitate solution is shaken with TCE or n-BuE vigorously, once shaking is stopped, the mixture immediately separates into a colloid precipitate phase and a clear solvent phase, implying that the solvents with functional groups –Cl and –O– cannot form stable H-bond with surface water molecules or surface –OH (see Table 1). When the mixtures are dried under an infrared lamp, grinding can disperse wet agglomerates into small particles and dried fine dust is obtained (see Fig. 2(H)), but the wet agglomerates become hard dried agglomerates without grinding (see Fig. 2(G)). The sedimentation experiments of colloid particles in other solvents have demonstrated that colloid particles can disperse well in these organic solvents with –OH or/and –COO–, suggesting that there are stable H-bonds between the functional groups and water molecules or –OH on the surface of the colloid.

| Compounds | *D.* (g/mL) | BP °C | Groups | Molecular Wt. (no groups) | Characteristics of dried product | | |
|---|---|---|---|---|---|---|---|
| | | | | | Color | Agglomerates | Hardness |
| Ethanol | | 78 | –OH | 29.06 | +++++ | +++++ | +++++ |
| 1,4-Butanediol | 1.016 | 235 | 2–OH | 56.10 | +++++ | +++++ | +++++ |
| EGMMeE | 0.964 | 124 | –OH,–O– | 43.09 | ++++ | +++++ | ++++ |
| TCE | 1.622 | 122 | 4 –Cl | 24.02 | +++ | ++++ | ++++ |
| n-Butanol | 0.810 | 117 | –OH | 57.11 | +++ | +++ | +++ |
| n-Pentanol | 0.815 | 138 | –OH | 71.14 | +++ | +++ | ++ |
| i-Amyl alcohol | | 132 | –OH | 71.14 | ++ | ++ | + |
| n-Butyl acetate | 0.881 | 126 | –COO– | 84.16 | + | ++ | + |
| n-Butyl ether | 0.769 | 142 | –O– | 114.23 | ++ | + | ++ |

a: Five plus indicates that colour is the deepest, agglomerates are the most, and hardness is the highest.

**Table 1.** Characters of products azeotropicly dried by different solvents [a].(Zhang. 2005)

The characteristics of the dried products prepared using n-pentanol, i-AmAl (i-Amyl alcohol), or n-BuA as an azeotropic agent have some differences, which, it is assumed, result from the differences in their molecular structures. Although compared with n-pentanol, i-AmAl is a more effective azeotropic agent and the obtained dried product is fluffier. However, the dried product treated by n-BuA is fluffier than that treated by i-AmAl. These results indicate that the improvement of fluffiness results from the molecular steric structure. n-BuA has the highest steric effect. Its two side chains stretch out and cover more surface area of colloid particles to prevent the particles from contacting each other. All of these suggested that the capacity of the azeotropic agent to remove water adsorbed on the surface of the colloid depends not only on its boiling point but also on its ability to replace the surface water molecules and indicated that the dispersivity and fluffiness of the dried product are controlled by the steric effect of the azeotropic agent. In this case, an effective azeotropic drying agent may be called an azeotropic dispersing agent.

The trends of dispersivity and fluffiness of the dried products changing with the functional groups and molecular structures of azeotropic solvents are shown in Fig. 3. Some empirical rules for selecting an azeotropic agent are drawn as follows: (1) the solvent molecule should contain at least one oxygen as the H-bond acceptor to form H bonds with the surface –OH of the polymer particle; (2) the H-bond acceptor should locate in the middle of the alkane chain rather than on the terminal so that the alkane can stretch out and cover more surface area; and (3) solvents should have a higher boiling point (~140ºC) to reduce the time of azeotropic distillation and the residual amount of azeotropic agent.

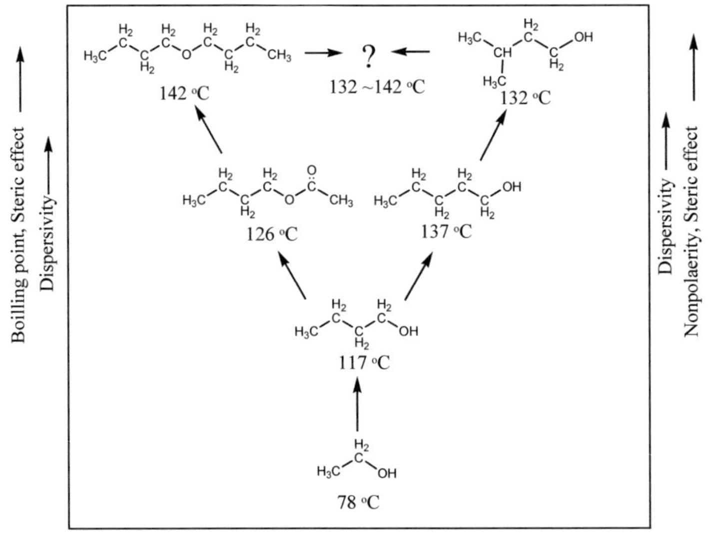


**Figure 3.** The variation of dispersivity and fluffiness of dried products with the groups and molecular structures of azeotropic solvents. (Zhang. 2005)

### *2.3.2. Iso-amyl acetate as an azeotropic agent (Zhang, 2005; Yang et al., 2007)*

The empirical rules and Fig. 3 guide us to find some other new organic azeotropic solvents for both drying and dispersing. Except for 2-hexanol, the other three organic solvents in Table 2 should have similar capacities of drying and dispersing colloid particles. The i-AmA (iso-Amyl acetate) is chosen as azeotropic solvent for it is commercially available and cost-effective among the three solvents. The behaviors of mixtures combining i-AmA with dense precipitate solution are monitored and measured carefully. (Yang et al., 2007).

The analysis of specific surface area of dried product is conducted for further substantiating the effects of i-AmA on the dispersivity. The adsorption isotherms are displayed in Fig. 4. The BET nitrogen surface areas are obtained by applying the BET equation to a relative pressure range of 0.05–0.3 on the adsorption isotherm. It can be seen that the BET surface area increases from 234.75 to 286.43 $m^2/g$ as the azeotropic solvent changes from the often-used n-butanol to i-AmA, and the dispersivity of the dried powder increases by 22%. Before the adsorption measurement, the dried product, large pieces of dried agglomerates, obtained with n-butanol has to be ground. However, the dried product from i-AmA is a highly dispersed powder such that it does not need further grinding, see the TEM images in Fig. 8. The IR spectrum of the dried powder derived from i-AmA shows lower residual organic compounds. XRD patterns and TEM images suggest a high mono-dispersivity of dried powder. These findings indicate that i-AmA is a much better azeotropic agent than other organic solvents.

| Solvent | Structure | Group | MMWt. | b.p.(°C) | *D*.(g/ml) |
|---|---|---|---|---|---|
| iso-Amyl acetate | 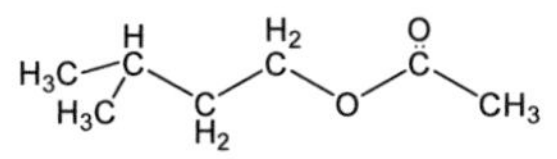 | -COO- | 98.19 | 143 | 0.88 |
| n-Amyl acetate | 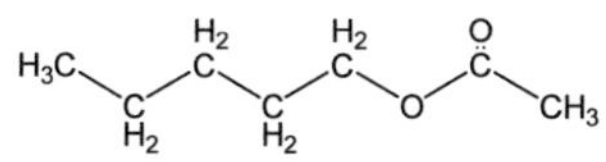 | -COO- | 98.19 | 149 | 0.88 |
| 2-Hexanol | 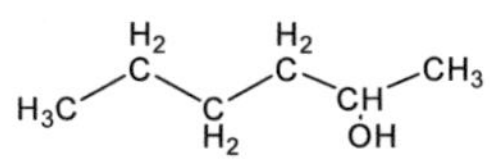 | -OH | 85.17 | 140 | 0.83 |
| Propyl butyrate | 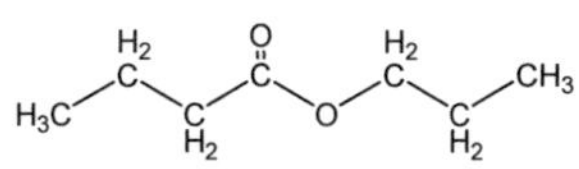 | -COO- | 98.19 | 142 | 0.88 |

**Table 2.** Structures and physical properties of the selected organic solvents. (Zhang. 2005)

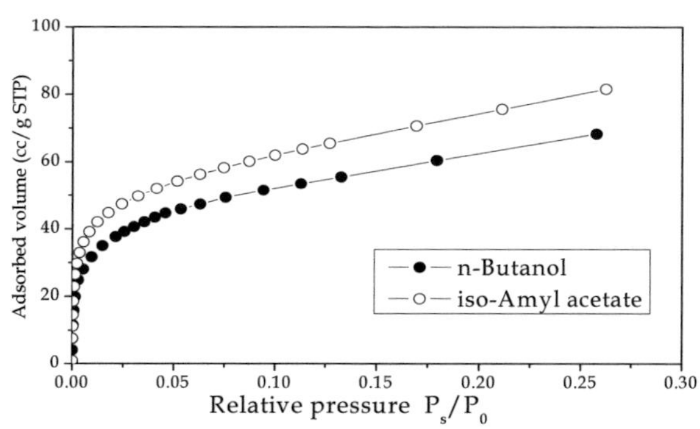


**Figure 4.** Nitrogen adsorption isotherms for dried Sb-doped $SnO_2$ powder on the SA3100 made by Becjman-Coulter co. USA. (Yang et al., 2007)

## 3. The formation of crystal particle in the process of synthesis (Zhang, 2008)

### 3.1. Functions of anion-exchange resin

The anion-exchange resin accelerates the hydrolysis of metal chloride, which is quite significant for the hydrolysis of the third- and fourth-order hydrolysis of $MCl_4$ and $NCl_3$ because the hydrolysis is reversible. However, if the removal of chlorine from hydrolyzate is carried out by washing with water or organic solvent, the hydrolysis will be a slow and a long-term process. When the number of group —OH borne by an atom M or N is more than two, condensation will take place between molecules by inter-molecularly dehydrating, or even gelatination occurs. The gelatination has the polycondensate form, a space network structure. The structure hinders the complete removal of fourth- or even third-order chlorine from gel and, therefore, can not lead to the formation of crystal of metal oxide because the remained chlorines block the condensation. See Scheme 2 A.

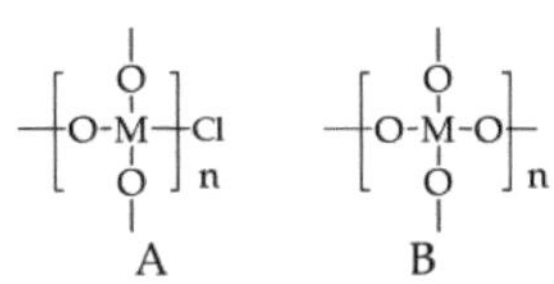


**Scheme 2.** The condensation of hydrolysis product.

The anion-exchange resin has a much higher affinity for $Cl^-$ than for $OH^-$, if once bonding a chloride ion to itself and a hydroxyl ion is given off to keep a constant concentration of $OH^-$. Due to anion-exchange resin can remove $Cl^-$ completely from solution, in this case, the hydrolysis product is high purity, which is of great advantage for the condensation of metal hydroxide to form crystal or assume a crystalline structure. See Scheme 2 B.

Another important function of ion-exchange resin is to crush masses of gel to nanoparticles like small balls in the Ball Grinder as the removal of chlorine is carried out on a shaker. In this way, the removal of chlorine from nanometer-sized particles is easy, fast, and efficient, which provides the condensation and crystallization with metal hydroxides bearing high purity. And also the function of the ion-exchange resin pellet analogous to that of ball milling results in a synthesized product with a nano-meter size and a very narrow size distribution.

Moreover, the removal of chlorine by ion-exchange resin instead of by repeated changing fresh solvent is of great importance to be used for accurate and repeated doping and for preparing nanocrystals with a narrow size distribution.

### 3.2. Functions of ethanol as solution medium

In the synthesis, the absolute ethanol is selected as a solution medium. Generally, metal salts, especially high valence metal salts, hydrolyzes evidently when they are dissolved in water, contributing to a acid reaction solution that cause part precipitation (segregation) of some metal salts as the doping is conducted. In addition, since a lot of metal salts have fast hydrolytic reaction rate, the use of water as solvent leads to an uncontrollable hydrolysis speed and a great deal of hydrolyzate settlement and agglomeration in a short time, which increases the difficulty of complete removal of chlorine from solid precipitates. A limited amount of water can control the hydrolytic rate under the ammonia as a controller of pH and the strongly anion exchange resin as a donor of $OH^{-}$.

The crystallization process in the solution synthesis of nanocrystal of metal oxide is essentially an inter-molecular dehydration process of metal hydroxides. A large quantity of water existing in the reaction system will definitely slow or even stop inter-molecularly dehydrating of the hydroxides. Ethanol, as been well known, is born with a strongly hygroscopic property and will pull water away from metal hydroxide to have it form oxide crystal. The water as a carrier of $OH^{-}$ between metal salt and ion-exchange resin, see reaction formula (8) and (9), is limited, whereas a large excess of ethanol plays an important role in promoting the crystallization by seizing water molecules from metal hydroxides, especially the promotion effect of ethanol is highly effective as the hydrolysis product is milled to nanometer size by ion-exchange resin balls on a shaker.

### 3.3. Behaviours of hydrolysis products and the crystal forming process in synthesis

The hydrolysis of many metal salts using the method of "P-CNAIE"follows some regularity. The stability of hydrolysis product colloid and the solution viscosity are related to the remains of chlorine in the hydrolysis solution, as shown in Fig. 5 and 6, which show the behaviours of the hydrolysate produced from $SnCl_4$ doped with $SbCl_3$ and from $ZnCl_2$ doped with $SbCl_3$. The left ones in both figures are the settlement ratio of hydrolysate in 2 h and right ones are the variation of the relative viscosity of solution containing hydrolysate.

We believed that the process of synthesizing nanocrystal undergoes three stages as shown in Fig. 7. At the first stage the sol, or linear molecule, is formed since where the sol is just the product of the first- and second-order hydrolysis of $MCl_4$ and $NCl_3$, which corresponds to the A to B in left figures of Fig. 5 and 6. The sol particles grow as the third- and fourth-order hydrolysis of $MCl_4$ and $NCl_3$ occurs and settle down at the bottom of the container at the second stage, corresponding to the line CD in the left figures in Fig. 5 and 6. Whereafter, the obtained precipitate is kept on reacting with ion-exchange resin to remove the remained chlorine and the precipitate re-suspends in ethanol solution at the third stage, as line EF in the left figures in Fig. 5 and 6. Here a further condensation takes place, leading to the contraction inward of particles and formation of nanocrystal in their central core.

Fig. 7 indicates that when the surface of precipitates bears a same material or electric charge (chlorine or water) precipitates become colloids suspending in solution. The phenomena are more evident for high valence metal than low valence metal. But for the alkali rare metal (such as Lanthanum) the hydrolyzate is unstable and readily settles down.

The variation of viscosity is evidence that substantiates the growth of sol in the synthesis of nanocrystal. The increase of the relative viscosity, $\eta_r$, of reaction solution shows the increase of the internal friction of hydrolysate, thus indicative of the increase of polymerization degree, or molecular weight, of hydrolysate. The successive ion exchanging does not add the molecular weight but cause the hydrolysate condensate inwards and internal friction decrease, contributing to relative viscosity drop. The inward condensation of particles accompanied by the dehydration between linear polymers is the crystallization process of colloid particles.

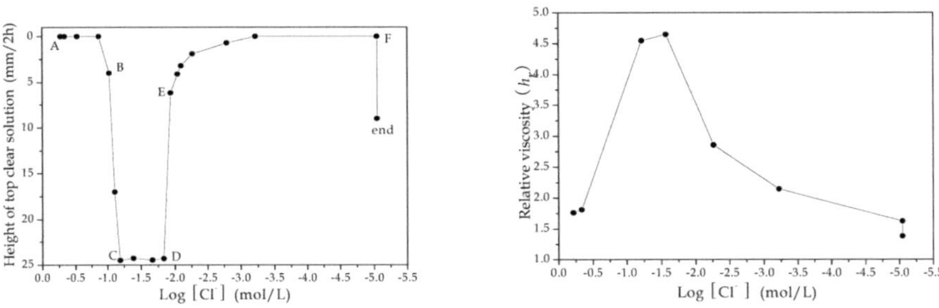


**Figure 5.** The stability of colloid in solution, see the left, and the solution relative viscosity, $\eta_r$, see the right, vary with $[Cl^-]$ in hydrolysis of $SnCl_4$ doped with $SbCl_3$. (The low limit of More Essay is $9\times10^{-6}$ mol/L for $Cl^-$).

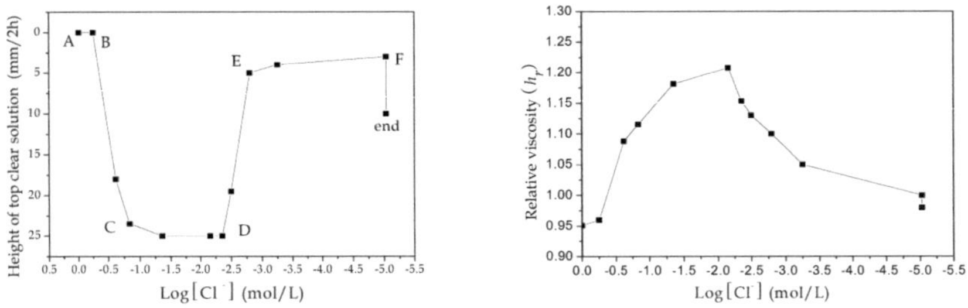


**Figure 6.** The stability of colloid in solution, see the left, and the solution relative viscosity, $\eta_r$, see the right, vary with $[Cl^-]$ in hydrolysis of $ZnCl_2$ doped with $SbCl_3$. (The low limit of More Essay is $9\times10^{-6}$ mol/L for $Cl^-$).

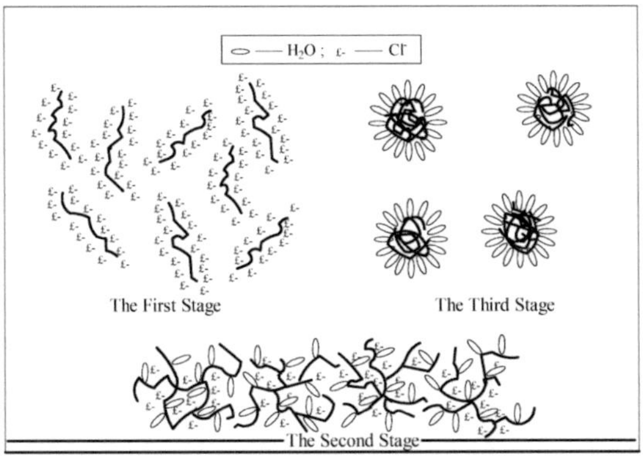


**Figure 7.** The three stages in the hydrolysis process of $MCl_4$ (or mixed with $NCl_3$).

## 4. Nanocrystal of stannic oxide doped with antimony

Nanocrystalline Sb-doped $SnO_2$ has many merits of stable chemical, mechanical, and optical properties and environmental stabilities. The introduction of the Sb element in the tin oxide lattice confers good conductivity to the materials and an extremely high potential of evolving oxygen in water. Moreover, the nanocrystalline Sb-doped $SnO_2$ has the potential to catalyze decomposition and oxidation for esters and alcohols. (Richards et al., 2001)

Binding energies of electrons and bandgaps between valence bands and conduction bands of materials are two significant parameters determining the properties of materials. Both are highly associated with the doping and the physical dimension of the material. The method of "P-CNAIE" can be used for accurate doping and to synthesize nanocrystals with a narrow size distribution because uniform and accurate doping at atomic or molecular levels confers stable and reliable properties and size of nanocrystals is associated with different quantum confinement effects (Yoffe 1993).

### 4.1. The synthesis

In an airtight flask containing 200 mL anion-exchange resin, 100 mL alcohol, and 10 mL of ammonia water, 200 mL ethanol solution containing $SnCl_4{\cdot}5H_2O$ (18.0%, w/v) and $SbCl_3$ (0.665%, w/v) are added dropwise with fast stirring. At the same time, $NH_3$ gas is aerated in the reaction solution (see Fig. 1 and Zhang et al. 2006). The reaction solution is held close to neutral by adjusting the speed of addition. After the addition is complete, a reaction solution is separated with resin through a glass-sand funnel and reactes repeatedly with 80 mL fresh anion-exchange resin on a shaker. The white precipitate gradually become light yellow during shaking. The final chlorine-free colloid solution is held idle on a bench for ca. ~48 h and separates into an upper lightly turbid solution and a lower dense precipitate. The upper lightly turbid solution is removed and kept aside for final recovery of all precipitate and the lower dense precipitate slurry is added ~80 mL of iso-amyl acetated to make a co-boiling system. A light-yellow dispersive fine powder is obtained by co-distilling off water absorbed on the colloid and solvents.

All the exchanged ion-exchange resins are combined and repeatedly washed with fresh solvent to collect any residual precipitate on the surface of the resin. The washed solvent is applied to a short column of ion exchange to remove any remaining chlorine, and is combined with the upper lightly turbid solution, in which, then, the resulting dried powder is dispersed on a shaker. In succession, the chlorine-free combined solution is distilled and leave behind a fine light-yellow powder. In this way, all of the metal hydrolysate can be recovered and an exact doping as experimenter desires is achieved.

### 4.2. The characterization of the structure

Fig. 8 shows TEM images and electron diffraction pattern of nano-meter sized material synthesized by the method "(P-CNAIE)" associating with "(AD-IAA)". The electro diffraction pattern in Fig. 8 indicates that the obtained nanomaterial has a determinate crystal structure, which is also confirmed by the TEM image B, from which a layer lattice structure can be distinctly identified.

In addition, XRD pattern in Fig. 8 illustrates the degree of crystallization and the size of nano particle. Diffraction peaks and their position in the pattern indicate the nano material is stannic oxide crystal, and broad and weak peaks suggest the size of the crystal is nanometer sized. The positions of peaks are consistent with that showed in the X-Ray Powder Diffraction Standards of $SnO_2$, PDF No. 41-1445 from Jade 5.0, see the red bar in Fig. 9. Fig. 10 is the TEM image of the finally fine powder treated at 650 °C for 3 h. It is quite obvious that the treatment at high temperature increases the size of nanoparticle in comparison with the size showed in Fig. 8A. Nevertheless, the dispresivity of both nanomaterials, unburned and burned, is almost the same.

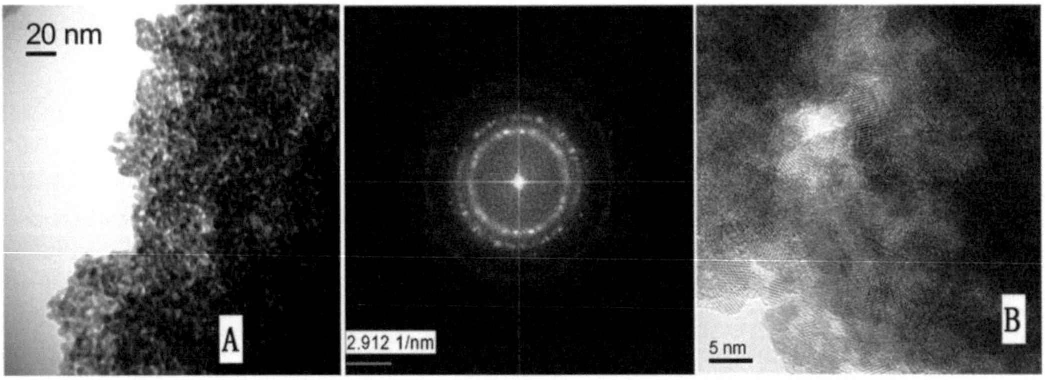


**Figure 8.** The TEM photoes of Sb-doped $SnO_2$ nanomaterial synthesized by method of P-CNAIE associate with AD-IAA and the EDP picture (JEM-2010FEF, JEOL, Japan). (Zhang, 2005)

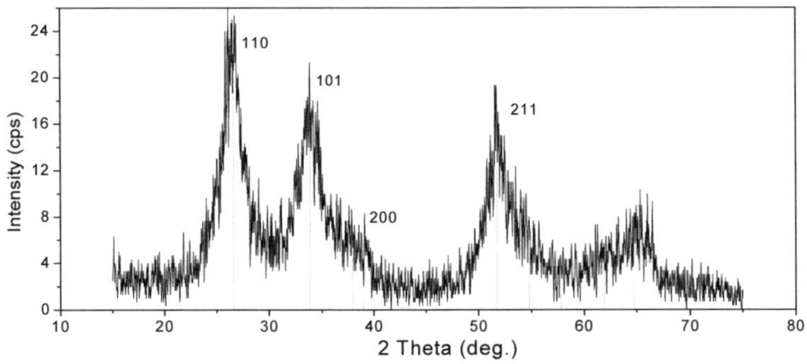


**Figure 9.** X-ray diffraction of Sb-$SnO_2$ nanomaterial synthesized by method of P-CNAIE associating with AD-IAA.

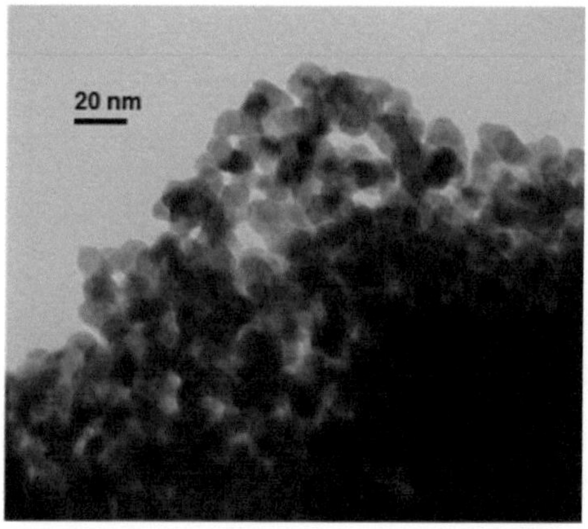


**Figure 10.** The fired Sb-$SnO_2$ nanomaterial synthesized by method of P-CNAIE associate with AD-IAA at 650°C. (Zhang et al., 2006)

## 4.3. The conductivity

### *4.3.1. Accuracy of doping*

To achieve a stable and reproducable conductivity of Sb-doped $SnO_2$, an accuracy of doping is critical. By the method of "P-CNAIE", it was observed that initial white precipitate gradually became finial yellow dried powder. Whereas, without doping, hydrolysate of pure stannic chloride from initial colloidal to finial dried powder is always white. These implicate: the yellow color is caused by Sb doping into condensate of stannic oxide, or, exactly, by Sb doping into crystal lattice of stannic oxide, because yellow is caused only by crystal with variation of band gap but by cluster or hydrolysate with forbidden band.

Because of the intermolecular condensation of hydroxides, the fine powder has no fixed chemical formula and, therefore, the total weight of the powder had to multiplied by a conversion factor, *r'*, which is a ratio of the conversion, associating with the crystallization, of hydroxide into oxide, see Section 6, in order to calculate the content of Sb in an oxide crystal. A fried antimony-doped stannic hydrolysate at >350 ºC does not dissolve in acid or basic solution and the method of fusing it with caustic soda results in serious errors, so the fresh dried powder is dissolved in HCl solution to make a sample for AAS. Fig. 11 shows a histogram of the content comparison of antimonies in powders and in reaction solutions before hydrolysis. The figure shows a high similarity between samples before and after the hydrolysis, within the molar fraction range of 0.040–0.075, and the highest relative error is only 1.24% (see Fig. 12).

It is not clear what causes the larger errors in the low-molar fraction range for Sb. It may be that the smaller low condensate of stannic hydroxide is more readily absorbed on the resin and difficult to be washed off with solvent. Or possibly, when being placed in the AAS, the sample was not atomized well, leading to an increase in the absorbing area of Sb and a high light absorption.

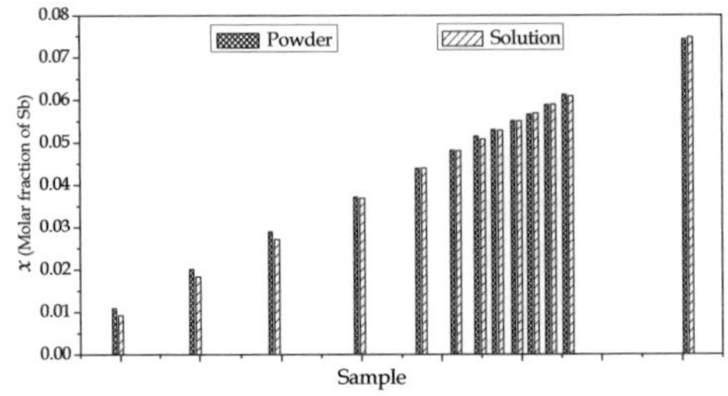


**Figure 11.** The comparison of Sb content in powders and in solutions before preparation reaction. The contents of Sb were detected on an atomic absorption spectrometer.

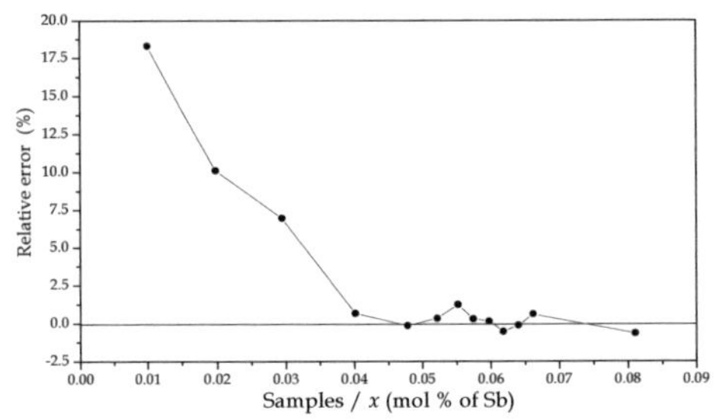


**Figure 12.** Relative errors in powder sample based on the content of solution before reacting.

Using anion-exchange resin to remove chlorine not only increased the recoverable yield but also ensured the accuracy and the similarity of doping, since the removal of chlorine is always carried out in the same solution. The only thing that needs to be replaced is the resin. Although there might been a small amount of precipitate absorbed on the surface of the resin, it can be washed off by shaking repeatedly with a fresh solvent and recovered by distilling off solvent. The worst aspect of removing chlorine by solvent washing is that the ratio of Sb to Sn is not fixed, whereas, in ion-exchange method, the proportion of Sb present in the solution before the reaction and in the powder are identical.

### 4.3.2. *Effect of doping on the conductivity*

$SnO_2$ is a semiconductor with a wide bandgap ($Eg$ = 3.97 eV) and transforms into a conductor after being doped with antimony (Gržeta et al., 2002). Our process provides a method to vary the content of Sb accurately. The ordinate in Fig. 13 indicates the logarithmic value of the resistivity of samples. There is an optimum ratio of Sb to Sn that gives the lowest resistivity, ranging between 5.28 and 5.50 mol% Sb, or 5.41 and 5.63 wt% Sb.

The variation of the resistivity of Sb-doped $SnO_2$ can be explained by the semiconductor fundamental theory. As the content of Sb increases and the density of the carrier and the carrier mobility, $\mu$, increase, the room-temperature resistivity of the crystal decreases. However, when the level of Sb is in excess of 5.50 mol%, as the carriers move in the semiconductor, the ionized impurity scattering can no longer be neglected. The increase in the number of scattering centers with addition of more antimony increases the amount of ionized impurity scattering and correspondingly decreases the carrier mobilities, which in turn increases the resistivity. Especially, the heavy doping will result in Burstein–Moss effects, which cause the bandgap to become wider (Cao et al., 1998; Irvin, 1962).

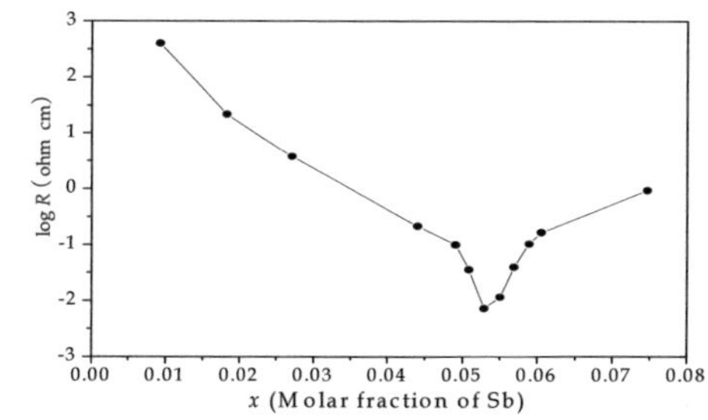


**Figure 13.** The variation of resistivity of Sb-doped $SnO_2$ with the content of Sb in it.

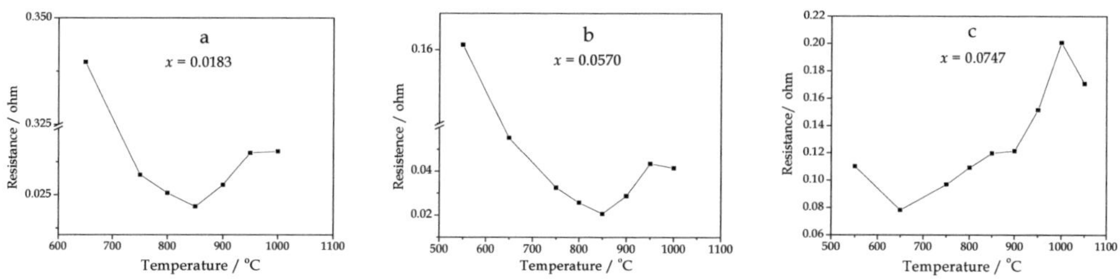


**Figure 14.** The variation of resistivity of nanocrystal with different content of Sb fired at different temperature.

Figure 14 shows the resistivity of Sb-doped $SnO_2$ as a function of temperature. Fine powders with identical Sb content were fired at different temperatures; materials with different resistivities were obtained (Castro & Aldao, 2000; Morikawa & Fujita, 2000).

It is proposed that the variation in resistivity with temperature is associated with oxidation of Sb on the surface layer of the nanocrystal. The doping of Sb in a low-temperature regime yields only $Sb^{+3}$, leading to p-type doping. However, Sb on the surface layer in a high-temperature regime may yield $Sb^{+5}$, leading to n-type doping. Crystals grown at high temperatures are more perfect, so the resistivity of a crystal decreases. However, above 850ºC, the number of $Sb^{+5}$ ions on the surface layer increases, so the surface layer is transformed into an n-type. Semiconductors do not conduct electricity when they are connected in an n–p–n connection. So with the temperature increases, the n-type layer becomes thicker and the crystallites gradually transform into poorer conductors. Figure 14 also shows that more $Sb^{+5}$ is formed at lower temperatures when the content of Sb is higher. Fig. 14 also indicates that the treatment at high temperature should be carefully adopted for the metal oxide with changeable valence.

### 4.4. Solid superacid of Sb-$SnO_2$ nanocrystal

Hino & Kobayashi indicated at first in 1979 that the acid strength of the $SO_4^{2-}/ZrO_2$ catalyst is estimated to be $H_0$ (Hammett indicator) ≥ -14.52, one of the strongest solid superacids. Sulfated zirconia ($SO_4^{2-}/ZrO_2$) is a typical solid superacid and exhibits a high catalytic activity for the skeletal isomerization of saturated hydrocarbons, and other reactions (Arata, 1990, 1996). Sulfated tin oxide ($SO_4^{2-}/SnO_2$) is one of the candidates with the strongest acidity on the surface. It has been reported that its acid strength is equal to that of $SO_4^{2-}/ZrO_2$ at least (Matsuhashi et al., 1989, 1990). Commonly, they have been prepared by the following procedures to generate superacidity (Matsuhashi et al., 2001): (i) preparation of amorphous metal oxide gels as precursors; (ii) treatment of the gels with sulfate ion by exposure to a $H_2SO_4$ solution or by impregnation with $(NH_4)_2SO_4$; (iii) calcination of the sulfated materials at a high temperature in air.

The sulfated Sb-doped $SnO_2$ crystal is prepared in our study as follows. 2 g of Sb-doped $SnO_2$ powder obtained in Section 4.1 is placed in a 50 mL plastic centrifuge tube containing 45 mL of methanol. After the powder is dispersed on a shaker, 3.0 mL of saturated ammonium sulfate, equal to ~2 g of $(NH_4)_2SO_4$, is added in methanol solution, and then the tube continues to be shaken on a shaker violently since once the saturated solution is dropped in methanol, very tiny $(NH_4)_2SO_4$ precipitate is separated in solution. After centrifugation, precipitate, a mixture of Sb-doped $SnO_2$ and $(NH_4)_2SO_4$, is washed in anhydrous alcohol and centrifuged at 4000r/min. The final sediment is dried under an infrared ray lamp and a dispersed powder is obtained.

DTA-TG curves in Fig. 15 exhibits the thermo-gravimetric turn point of Sb-doped $SnO_2$ powder without mixing $(NH_4)_2SO_4$ is 376°C. Differential scanning calorimetry pattern shows the giving out heat peak is 341°C.

The relative acid strength of the powder was measured by the adsorption reaction of indicator. The powder (ca. 0.5 g) is calcined at 613 K in air for 3 h and then placed in a glass vacuum desiccator as the powder is hot. After the sample is pretreated in a vacuum for 2 h and cooled down to room temperature, some cyclohexane solution containing 5% of Hammett indicator is sucked in the vacuum desiccator. The desiccator is heated to 60 °C by placing it in a constant water bath, which results in the powder exposing to indicator vapor. The present powder sample is gradually colored by indicator and changes distinctly the colorless basic form of p-nitrotoluene ($pK_a$ or $H_0$ = -11.4), m-nitrotoluene (-12.0), m-nitrochlorobenzene (-13.2), and 2,4-dinitrotoluene (-13.8) to the yellow conjugate acid form and slightly yellow of 2,4-dinitrofluorobenzene (-14.5), that is to say, the acid strength of the solid acid is estimated at least to be $H_o \approx -14.5$.

In the study, it is found that superficial hydroxyl on Sb-$SnO_2$ nanoparticles is very important for the sulfating roasting of Sb-$SnO_2$ nanocrystal. The fewer the number of superficial hydroxyl exist, the fewer the sulfated groups exist on the surface of Sb-$SnO_2$ nanocrystals. Compared with curves 1 (Sb-$SnO_2$) and 8 $(NH_4)_2SO_4$, curves 2 to 7 (Sb-$SnO_2$ + $(NH_4)_2SO_4$) have an additional line segment from **a** to **b**. It is easy to understand that the line segment implies the generation of superficially sulfated groups. Curves 2 to 7 are thermogravimetric curves of Sb-$SnO_2$ nanocrystals that are pretreated at different temperature for 3 h and then impregnated with a given amount of $(NH_4)_2SO_4$ before thermo-gravimetric analysis. It can be seen that the higher the preprocessing temperatures is, the shorter the line segments from **a** to **b** and the fewer the amount of sulfated group is. As the preprocessing temperature increases, especially at or over 341 °C, the weight losses of Sb-$SnO_2$ nanocrystals is heavier, which is contributed by the dehydration between hydroxyls, leading to dwindling in the number of superficial hydroxyls that are able to bond sulfate radicals and in the number of superficially sulfated groups and to decreasing acid strength or catalytic activities.

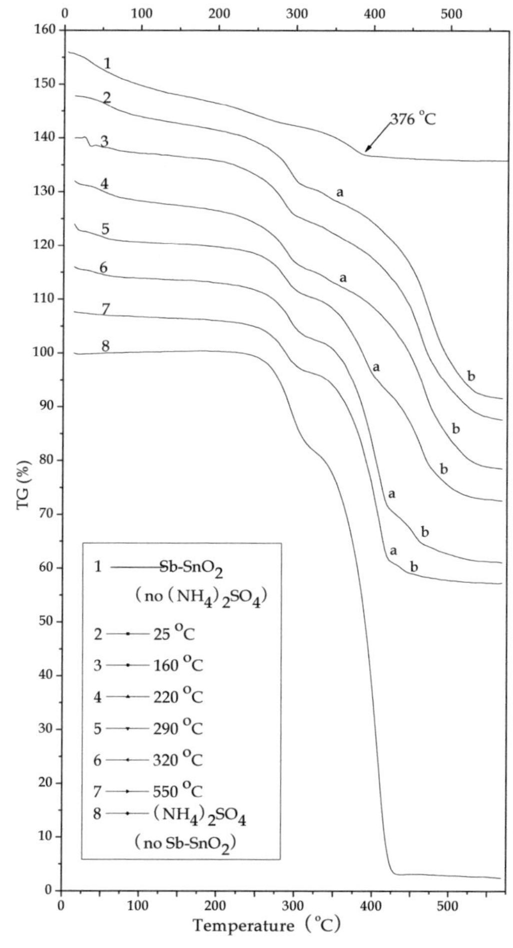


**Figure 15.** Thermogravimetric curves of Sb-$SnO_2$, Sb-$SnO_2$ + $(NH_4)_2SO_4$, and $(NH_4)_2SO_4$.

Fig. 15 indicates that due to the generation of sulfated $SnO_2$, the binding of $SO_4^{2-}$ on the $SnO_2$ endows the group with additional energy, so that to decompose or to volatilize the bonded $SO_4^{2-}$, an excessive energy has to be provided in order to overcome the bonding force. Therefore line segments **a** to **b** appear in TG curves.

Fig. 15 substantiates the advantage that nanocrystal synthesized by the solution method has over the calcined nanocrystal in the surface modification of nanomaterials due to the existence of superficial hydroxyls.

## 5. Nanocrystal of zinc oxide and zinc oxide doped antimony or magnesium (Zhang et al., 2011, Liu et al., 2008)

Inorganic antimicrobial is a widely used as antibacterial material, which makes use of the blocking effect of metal ions and metal oxide to disinfect and decontaminate. Silver, as an inorganic antimicrobial, is widely known for its antibacterial properties against most of bacteria species. However, metal oxides, with higher stability in physics and chemistry and wide-spectrum and persistent antibacterial properties, show a high effective bactericidal potency at common circumstance. In particular, nano-sized inorganic antimicrobial expands the range of bactericidal application. Most inorganic antimicrobials are photocatalytic material，such as semiconductor：$TiO_2$, ZnO, $Fe_2O_3$, $WO_3$ and CdS etc. These materials can act with water under irradiation to produce non-selective and highly active hydroxyl radicals (·OH) that can kill or inhibit proliferation of microbe (Linsebiglar, et al., 1995, Stevenson, et al., 1997). Of them ZnO is an economic, resourceful, chemically stable and non-toxic photocatalytic antibacterial. Due to the low thermal expansion and cold contraction coefficients and highly chemical stability, zinc oxide is used in industrial and common living ceramics and also added to coating and paints.

### 5.1. Synthesis of nanocrystal of doped zinc oxide

In a flask containing 400 mL ethanol solution of a desired ratio of molar concentrations of $ZnCl_2$ to $SbCl_3$ or $MgCl_2$ and 200 mL of anion exchange resin, 5 ~ 20 mL double-distilled water are added dropwise under stirring. The reaction is then conducted on a shaker and the solution containing suspended precipitate is separated with resin by a 120 mesh strainer. The filtrate containing precipitate reacts repeatedly five or six times with 50 mL fresh anion-exchange resin on a shaker until the upper solution does not become clear upon short standing (~ 2 h). The final chlorine-free solution is held idle on a bench for ca. 48 h. The upper lightly turbid solution is removed and kept aside. Iso-amyl acetate is added to the lower dense precipitate slurry. After mixed on a shaker for 120 min the mixture is dried in a glass distillation apparatus and a highly dispersive fine powder was obtained. All the reacted ion-exchange resins are combined and repeatedly washed with fresh solvent. The washed solvent is applied to a short column of ion exchange to remove any remaining chlorine, and is then combined with the upper lightly turbid solution. The dried powder from azeotropic distillation is dispersed in the combined solution on a shaker, and then distilled off the solvent and leave behind the fine powder with exactly dopant content.

### 5.2. The crystal structure of zinc oxide

The nanoparticle of Sb-ZnO and Mg-ZnO synthesized by the method of P-CNAIE associating with AD-IAA has a crystal structure even without calcining.

But it is discovered from Figure 16 that, when the doping level is higher than 1/10 (mole ratio), the crystal lattice of zinc oxide undergoes significant change. The XRD patterns of doping ratio at 1:5 for Mg-doped and Sb-doped ZnO crystals and 1:10 for Sb-doped ZnO indicate the formation of a new crystal phase. Moreover, Fig. 17, an enlarged figure, shows that the new crystal phase caused by Sb doping is different from that by Mg doping in crystal structure, which might be one of major causes that lead to a difference in bacteriostatic potency between Mg-doped and Sb-doped ZnO nanocrystals. The Mg doping contributed to 5 new XRD peaks appear at 17.0°, 22.4°, 25.5°, 28.6° and 38.3 °(2θ), while the Sb doping brought on 3 new peaks at 17.0°, 24.6° and 38.3° (2θ). (See Fig. 17).

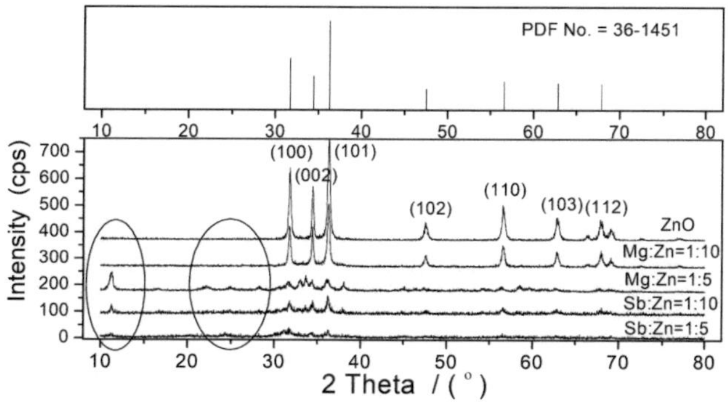


**Figure 16.** XRD patterns of ZnO nanoparticle doped with $Sb^{3+}$ or $Mg^{2+}$. The up patterns is the X-ray powder diffraction standards, PDF No. 36-1451, of ZnO from Jade 5.0.

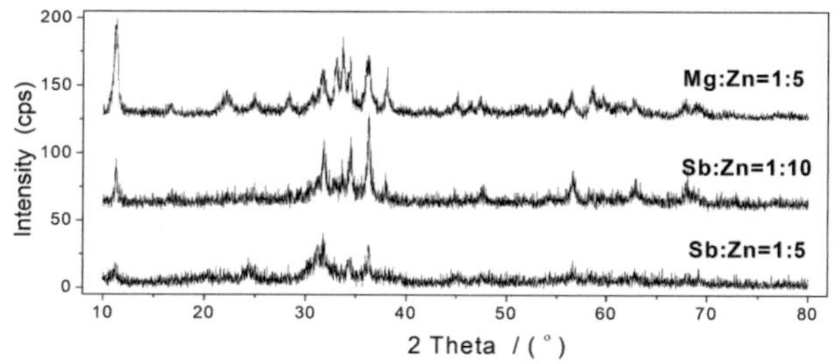


**Figure 17.** XRD patterns of ZnO nano-particle doped with $Sb^{3+}$ or $Mg^{2+}$ at heavy doping.

Variation of the structure of doped ZnO can be seen under an electron microscope. TMS images indict such variation induced by doping Mg and Sb. Fig. 18A and E show non-doped ZnO with a structure of typical columnar crystal and bigger size. In the case of doping, the size of Sb-doped ZnO crystal evidently decreased with Sb content increasing, as Fig. 18B, C and D show. It can be seen from Fig. 18C and D that besides many short columnar crystal a great deal is grained crystal. However, as Mg is doped into ZnO, the crystal size does not change a lot, but the crystalline form varies from six-rowed columnar to six-rowed lamellar. The lamellar crystal appears in very small amount as doping level is low (see Fig. 18E, indicted by arrow), and in a great amount as doping ratio reached to 1 : 5 (see Fig. 18F).

The outer-shell valence electrons of zinc atoms are four (2s and 2d) and its ionic radius 0.60 Å, the valence electrons of magnesium are two (2s) and ionic radius only 0.57 Å, while that of antimony are five (2s and 3p) and the ionic radius up to 0.74 Å. Therefore, when the doping is heavy, the effects of differences in the number of outer-shell electrons and in the ionic radius on the variation in crystal form are more distinct, contributing to the transformation of original six-rowed columnar crystals into new crystal form.

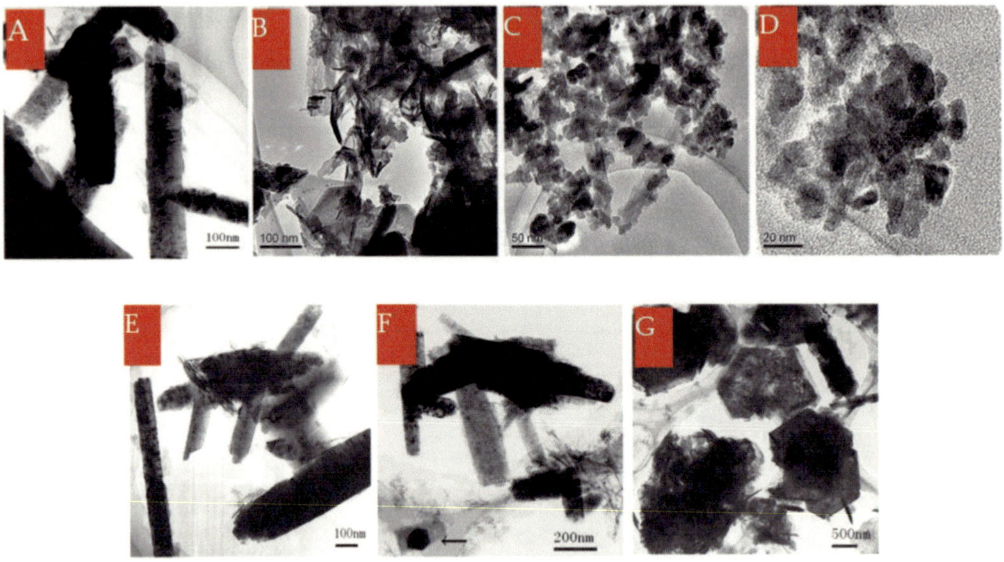


**Figure 18.** TEM (JEM-2010 (HT), JEOL, Japan) images of doped and non-doped zinc oxides. A is a pure ZnO, B: Zn : Sb = 1:15, C: Zn : Sb = 1:10 and D: Zn : Sb = 1:5. The low group, E is a pure ZnO, F: Mg : Zn = 1:10 and G: Mg : Zn = 1:5.

When element Mg(II) is doped into matrix element Zn(II), both of them have the same valence and the same ratios of metal atom to oxygen atom, which will not cause any hanging bonds within crystal, but the distortion of lettice resulted from the difference in ionic radius between matrix and doping elements at heavy doping lead to the formation of hanging bond and hence an enhancement of photo-catalysis of the doped crystal. Whereas in the case of Sb doping, Sb(III) has one more oxygen atom aroud it than Zn(II), which causes a hanging bond on Sb atom, or a unbonding valence electron. This unbonding electron becomes a free electron within crystal lattice, resulting in the number of free electron more than the number of positive hole and becoming a n-type semiconductor. As is well known, ZnO crystal is a n-type semiconductor. Consequently, Sb-doping promotes the semiconductor charactristic of ZnO, endowing it a more effective photo-catalysis. In addition, the difference in ionic radius also assisted the distortion of crystal lettice, so the variation of ZnO crystal as antimony dopes is more obvious than as magnium dopes.

A relevant doping, as at 1/10, is important for to maintain ZnO's semiconductive nature as well as its basic columnar structure. The main change is the particle size (Fig. 18C and the arrow pointed in F). Such doping provides the nanomaterial an enhanced semiconductive characteristic and effective photo-catalysis. The formation of new crystal, as XRD peak at 17.0° shows, does not much help to increase the photo catalysis, or bactericidal potency because, we supposed, the bactericidal potencies at doping ratio of over 1/5 are much lower than these at doping ratio of 1/10. Maybe the new crystal is not a semiconductor and hence can not produce hydroxyl free radical when it acts with water at irradiating.

### 5.3. The bactericidal properties

The bacteriostatic rate of Sb-doped ZnO nano powder is only 12% as the plats are incubated in darkness. While check experiments of Sb-doped ZnO powder are carried out under lighting, which leads to a bactericidal rate up to 93.4%, increasing by over 80%. The Mg-doped ZnO powder improves its bacteriostatic rate from 9.8% without irradiating to 83.5%. Fig. 19 shows that the doped materials have a much high antibacterial performance than non-doped materials at both irradiating and darkness conditions. Among tested nanomaterials, Sb-doped crystal presents the highest bactericidal potency, while the potency from non-doped material is the lowest. Meanwhile, the irradiation enhances antibacterial rate of all tested nano crystals. The light does play a vital role on the antibacterial behaviors of tested nanomaterials, because of the limited antimicrobial properties in darkness. We supposed that such antibacterial behaviors are due to the reactions between nanocrystal and water molecule under visible-light irradiation to produce free radical. The antibacterial behavior caused by irradiating disappears once the irradiation is removed. The limited bactericidal potency under darkness might be caused by the limited amount of free radical that is produced by powder absorbing environmental energy to act with water molecule.

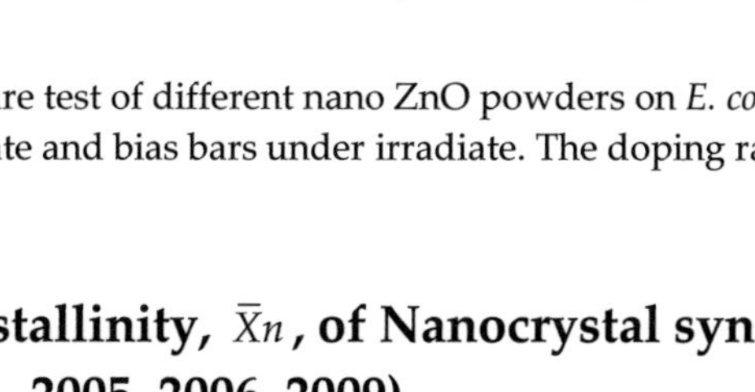


**Figure 19.** The lighting culture test of different nano ZnO powders on *E. coli*. Dark bars show the incubation under non-irradiate and bias bars under irradiate. The doping ratio of Mg-doped ZnO and Sb-doped ZnO are 1/10.

## 6. Calculation of crystallinity, $\bar{X}n$, of Nanocrystal synthesized by solution method (Zhang et al., 2005, 2006, 2009)

Metal hydroxide can be burned at high temperature, and water loss has it transformed into an oxidate. So the remained metal hydroxide that dose not condensate to form oxidate crystal can be calculated by weighing water loss as the superficial metal hydroxide is burned. A simple and reliable method to determine the number of water molecules, a by-product in the polycondensation of metal hydroxide into metal oxide crystal, is proposed in order to calculate the average crystallinity of nanocrystal synthesized by solution method. The number of water molecules can be determined by weighing the difference between synthesized nanocrystals before and after burning. There is a close circle relationship (tin in this calculation example), by which the average crystallinity can be calculated.

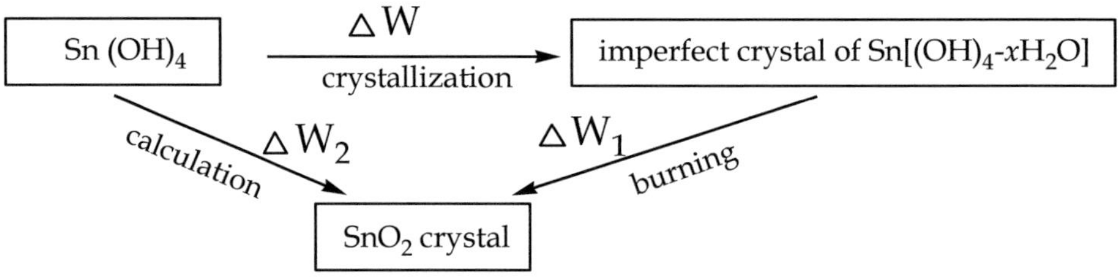


**Scheme 3.** The conversion relationship of monomer, polycondensate and crystal, and the cyclic relationship of calculation of water molecule numbers.

$$\text{No. of water molecules} = \frac{\Delta W}{M_{H_2O}} = \frac{\Delta W_2 - \Delta W_1}{M_{H_2O}}$$

Here, $Sn(OH)_4$ and $SnO_2$ have definite molecular weights and can be calculated or experimentally determined. So the number of water molecules produced as condensation, or crystallization, is given. Being divided by a parameter $n_0$, which is the number of tin atom or monomer $Sn(OH)_4$, the equation changes as follow and the $n_0$ can be described as Eq. (10),

$$\frac{(\text{No. of water molecules}) \times M_{H_2O}}{n_0} = \frac{W_{Sn(OH)_4}}{n_0} - \frac{W_{polycondensate}}{n_0}$$

$$n_0 = \frac{(\text{No. of water molecules}) \times M_{H_2O}}{M_{Sn(OH)_4} - M_{polycondensate}} \tag{10}$$

Here $W_{polycondensate}$ stands for $W_{\text{imperfect crystal of } Sn[(OH)4\text{-}xH2O]}$ or $W_{Sn[(OH)4\text{-}xH2O]}$. And the number of end groups of the polycondensate molecules is given by

$$\text{No. of end hydroxyl groups of imperfect crystal} = 2 \times \frac{\Delta W_1}{M_{H_2O}}$$

The extent of conversion, $r'$, that is, polycondensate, or imperfect crystal converts into perfect crystal of $SnO_2$:

$$r' = W^*_{SnO_2} \big/ W^*_{polycondensate} = n_0 \times M_{SnO_2} \big/ n_0 \times M_{Sn[(OH)_4 - xH_2O]}$$

And so

$$M_{Sn[(OH_4\text{-}xH_2O]} = M_{SnO_2} / r' \tag{11}$$

where W $*_{polycondensate}$ and W $*_{SnO2}$ are the masses of the imperfect crystal to be fired and of the oxide formed after firing, respectively. The extent of theoretical conversion is:

$$W_{SnO_2} / W_{Sn(OH)_4} = M_{SnO_2} / M_{Sn(OH)_4} = R < r'$$

because the condensation results from intermolecular elimination of water and hence $M_{Sn(OH)4} > M_{Sn[(OH)4\text{-}xH2O]}$. Substituting Eq. (11) for $M_{Sn[(OH)4\text{-}xH2O]}$ in the expression Eq.(10), a general expression is given as Eq. (12)

$$n_0 = \frac{(\text{No. of water molecules}) \times M_{H_2O}}{M_{Sn(OH)_4} - \frac{M_{SnO_2}}{r'}} \tag{12}$$

Once the number of water, or the number of condensation reactions, is given, the $n_0$ can be calculated using $r'$. The $n_0$ will only be the function of $r'$. With $r'$ increases, the number of initially presented monomer, $n_0$, should be less, meaning the degree of condensation increase. For convenience's sake, the number of condensation reactions, or No. of water molecules, can be factitiously set on requirement of calculation accuracy, say 100, 1,000, 10,000 and so on. Its significance is that in each $n_0$ of monomers, intermolecular condensations occur by the elimination of a certain water molecules between monomers. The definition of the $n_0$ is the number of initially presented monomer or the total molar number of metal elements.

**For example:** In the preparation of Sb-doped $SnO_2$, W $*_{Sn(OH)\,x}$ = 0.9573 g, W $*_{SnO2}$=0.8836 g and so r′ is 0.9320. In addition, M $_{Sn(OH)4}$=186.71, M $_{SnO2}$=150.71, M $_{H2O}$=18.002. So the Eq. (12) is written as follows when the No. of water molecules is set as 1,000.

$$n_0 = \frac{18.002 \times 1000}{186.71 - \frac{150.71}{0.9230}} = 768$$

It indicates that 768 of hydroxides take place 1,000 condensation reactions by losing 1,000 water molecules. According to the definitions of the average degree of polymerization, $\overline{X}_n$, which here should be called as the average crystallinity, the expression is

$$\overline{X}_n = \frac{\text{No. of units}}{\text{No. of polymers}} = \frac{N_0}{N}$$

For linear condensation, including side-chain condensation, the average effective functionality (i.e. the number of condensable functional groups) of every monomer is supposed to be two, i.e. $\overline{f} = 2$, and the equation of Carothers can be directly used to calculate the average degree of polymerization, $\overline{X}_n$.

$$p = (N_0 - N)/N_0 = 1{,}000/n_0$$

$$\overline{X}_n = 1/(1-p) = n_0/(n_0 - 1{,}000) \tag{13}$$

where $p$ is the extent of reaction, and $N_0$ and $N$ are respectively the total numbers of molecules initially and finally present, respectively. $(N_0\text{–}N) = 1{,}000$ (water molecules) and $N_0 = n_0$. For $n_0 = 2000$, $\overline{X}_n = 2$ ; $n_0 = 1125$, $\overline{X}_n = 9$ and $n_0 = 1000$, $\overline{X}_n = \infty$. See the $\overline{X}_n$ in Table 3. For nonlinear condensation, products have a network structure, which, conceivably, might be a planar network analogous to that of graphite, or a space network analogous to that of diamond. Compared and analyzed a great number of typical structural units, values of $\overline{X}_n$ in this case can be obtained using expressions,

$$N = 2 \times 1000 / (\overline{X}_n \overline{f} - \sqrt{\overline{X}_n \overline{f}})$$

$$\overline{X}_n = N_0 / N = \left[N_0\overline{f} / (N_0\overline{f} - 2\times 1000)\right]^2 = \left[n_0\overline{f} / (n_0\overline{f} - 2\times 1000)\right]^2 \tag{14}$$

where $\overline{X}_n \overline{f}$ refers to the total number of functional groups before condensation, $\sqrt{\overline{X}_n \overline{f}}$ is the number of unreacted groups remaining within the structure units after condensation, and $2\times1000$ is the number of reacted groups. As $n_0 = 1000$, $\overline{X}_n = 4$; $n_0 = 750$, $\overline{X}_n = 9$; $n_0 = 667$, $\overline{X}_n = 16$; and $n_0 = 500$, $\overline{X}_n = \infty$.

| $r'$ (%) | $n_0 \times 1000$ | $\overline{X}_{n(linear)}$ † | $\overline{X}_{n(net-linear)}$ † | $\overline{X}_{n(space-net)}$ † |
|---|---|---|---|---|
| 80.72 | ∞ | Single | Single | Single |
| 80.75 | 225025 | | | |
| 80.8 | 94747 | | | |
| 81 | 27695 | | | |
| 82 | 6165 | | | |
| 84 | 2651 | | | |
| 84.80 | 2000 | 2 | | |
| 87.46 | 1250 | 5 | | |
| 88.28 | 1125 | 9 | | |
| 88.77 | 1063 | 17 | | |
| 89.14 | 1020 | 50 | | |
| 89.33 | 1000 | ∞ | 2 | 4 |
| 92.63 | 750 | | 3 | 9 |
| 94.36 | 667 | | 4 | 16 |
| 97.43 | 562 | | 9 | 81 |
| 98.54 | 533 | | 16 | 256 |
| 99.54 | 509 | | 49 | 2401 |
| 100 | 500 | | ∞ | ∞ |

†: Assuming only one kind of structural unit is yielded in the condensation.

**Table 3.** The $\overline{X}_n$ , crystallinity, corresponding with $r'$and $n_0$ .

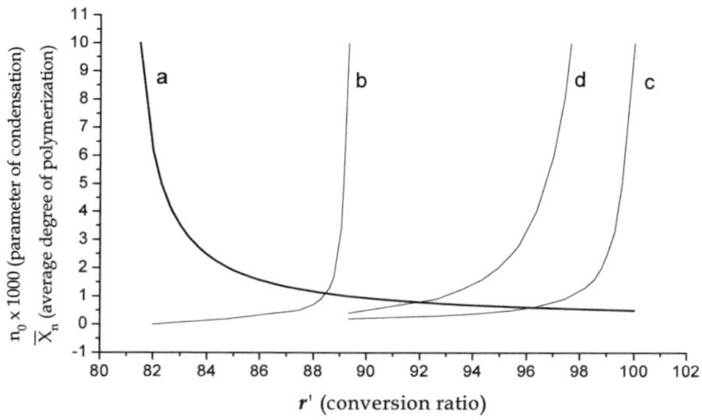


**Figure 20.** In a given condensation reaction, the relation, denoted as a, between $n_0$ and $r'$ according to Eq. (12), and the relations, denoted as b, c and d, between $\overline{X}_n$ and $r'$ according to Eqs. (13) and (14), respectively.

## Author details

Xuejun Zhang
*School of Chemistry and Chemical Engineering, Guizhou University, Guiyang, Guizhou, China*

Fuxing Gan
*School of Resource and Environmental Science, Wuhan University, Wuhan, Hubei, China*

## Acknowledgement

The authors thank Dr. Fen Yang, Dr. Hua Zhu, Dr. Xuhui Ma, Peng Liu, Liye Fu, Rong Zhou, Hanmei Ouyang and Hui Zhong for participating in parts of experimental work, and give many thanks to Prof. Chris Abell of the Chemistry Department, University of Cambridge, for the correction on part of manuscript. We are extremely thankful to the "Chun Hui" Project of the Ministry of Education of China for funding this research.

## 7. References

Arata, K. (1990). Solid Superacides. *Adv. Catal.* Vol.*37*, pp. 165-211. ISSN: 0360-0564

Arata, K. (1996). Preparation of superacids by metal oxides for reactions of butanes and pentanes. *Appl. Catal, A: General*, No.146, pp. 3-32.

Brinker, C. J. & Scherer, G. W. (1990). *Sol–Gel Science: The Physics and Chemistry of Sol–Gel Processing*, pp. 89–103. Academic Press, Boston, and references therein.

Boujday, S. Wünch, F. Portes, P. Bocquet, J-F. and Christophe, C-J. (2004). Photocatalytic and Electronic Properties of TiO2 Powder Elaborated by Sol–Gel Route and Supercritical Drying. *Sol. Energy Mater. Sol. Cells*, Vol.83, No.4, pp. 421–33.

Cao, X. Cao, L. Yao, W. & Ye, X. (1998). Influence of Dopant on the Electronic Structure of SnO2 Thin Films. *Thin Solid Films*, Vol.317, No.2, pp. 443–5.

Castro, M. S. & Aldao, C. M. (2000). Effects of Thermal Treatments on the Conductance of Stannic Oxide. *J. Eur. Ceram. Soc.*, Vol.20, No.3, pp. 303–7.

Diao, Y. Walawender, W. P. M. Sorensen, C. Klabunde, K. J. & Ricker, T. (2002). Hydrolysis of Magnesium Methoxide. Effects of Toluene on Gel Structure and Gel Chemistry. *Chem. Mater.*, Vol.14, No.1, pp. 362–8.

D'Souza, L. Suchopar, A. & Richards, R. M. (2004). In Situ Approaches to Establish Colloidal Growth Kinetics. *J. Colloid Interf. Sci.*, No.279, pp.458–63.

Frazee, J. & Harris, T. (2001). Processing of Alumina Low-Density Xerogels by Ambient Pressure Drying. *J. Non-Cry. Sol.*, Vol.285, No.1–3, pp. 84–9.

Gržeta B. Tkalčec E. Goebbert C. Takeda M. Takahasgi M. Nomura K. & Jaksic M. (2002). Structural Studies of Nanocrystalline $SnO_2$ Doped with Antimony: XRD and Mössbauer Spectroscopy. *J. Phys. Chem. Solids,* Vol.63, No.2, pp. 765-72

Hino, M. Kobayashi, S. & Arata, K. (1979). Reactions of Butane and Isobutane Catalyzed by Zirconium Oxide Treated with Sulfate Ion. Solid Superacid Catalyst, *J. Am. Chem. Soc.* No.101, pp. 6439-6441

Hu, Z. Dong, J. & Chen, G. (1998). Replacing Solvent Drying Technique for Nanometer Particle Preparation. *J. Colloid Interface Sci.*, Vol.208, No.2, pp. 367–72.

Hwang, K.-T. Auh, K.-H. Kim, C.-S. Cheong, D.-S. & Niihara, K. (1997). Influence of SiC Particle Size and Drying Method on Mechanical Properties and Microstructure of $Si_3N_4$/SiC Nanocomposite. *Mater. Lett.*, Vol.32, No.4, pp. 251–7.

Irvin, J. C. (1962). Resistivity of Bulk Silicon and of Diffused Layers in Silicon. *Bell Systems Tech. J.*, No.41, pp. 387–91.

Linsebigler, A. Lu, G. & Yates, J. Jr. (1995), Photocatalysis on $TiO_2$ surfaces-principle, mechanisms, and selected results, *Chem Rev*, Vol95, No.3, pp. 735-758.

Liu, P. Guo, Z. Zhang, X. & Xiao, T. (2008). Novel Anion Exchange Method for Synthesis of Zinc Oxide Nanocrystal Powder and Photocatalytic Activity of the Powder, *Nanoscience & Nanotechnology* (China), Vol.5, No.1, pp. 64-69.

Luliang, Q. Yongcun, L. & Yang, X. (2000). *Handbook of Chemicals and Materials for Treatment of Water*. p. 217. China Petrol and Chemistry Publishing House, Beijing.

Luan, W.-L. Gao, L. & Guo, J.-K. (1998). Study on Drying Stage of Nanoscale Powder Preparation. *Nanostructured Mater.*, Vol.10, No.7, pp. 1119–25.

Matsuhashi, H. Hino, M. & Arata, K. (1989). In Acid-Base Catalysis; Tanabe, K., Hattori, H., Yamaguchi, T., Tanaka, T., Ed. p 357. Kodansha: Tokyo.

Matsuhashi, H. Hino, M. Arata, K. (1990). Solid catalyst treated with anion: XIX. Synthesis of the solid superacid catalyst of tin oxide treated with sulfate ion. *Appl. Catal.* Vol.59, No.1, pp. 205-212.

Matsuhashi, H. Miyazaki, H. Kawamura, Y. Nakamura, H. and Arata, K. (2001). Preparation of a Solid Superacid of Sulfated Tin Oxide with Acidity Higher Than That of Sulfated Zirconia and Its Applications to Aldol Condensation and Benzoylation1, *Chem. Mater.* No.13, pp. 3038-3042

Morikawa, H. & Fujita, M. (2000). Crystallization and Electrical Property Change on the Annealing of Amorphous Indium-Oxide and Indium–Tin-Oxide Thin Films. *Thin Solid Films*, Vol.359, No.1, pp. 61–7.

Park, C. Bell, A. T. & Tilley, T.T.D. (2002). Oxidative Dehydrogenation of Propane Over Vanadia–Magnesia Catalysts Prepared by Thermolysis of $OV(OtBu)_3$ in the Presence of Nanocrystalline MgO. *J. Catal.*, Vol.206, No.1, pp. 49–59.

Richards, R. & Khaleel, A. (2001). Ceramics, in: *Nanoscale Materials in Chemistry*, K. J. Klabunde, (Ed), pp. 95–9, Wiley VCH, New York.

Richards, R. Mulukutla, R. Volodin, A. Zaikovski, V. Sun, N. & Klabunde, K. (2001) Nanocrystalline Ultra High Surface Area Magnesium Oxide as a Selective Base Catalyst. *Scripta Materialia*, Vol.44, No.8–9, pp. 1663–6.

Shlyakhtin, O. Oh, Y-J. & Tretyakov, Y. (2000). Preparation of Dense $La_{0.7}Ca_{0.3}MnO_3$ Ceramics from Freeze-Dried Precursors. *J. Eur. Ceram. Soc.*, Vol.20, No.12, pp.2047–54.

Stevenson, M. Bullock, K. & Lin, W. (1997). Sonolytic enbancement of the bactericidal activity of irradiated titanium dioxide suspensions in water. *Res Chem Intermed*, Vol.23, No.4, pp. 311-323.

Sun, C. and Berg, J. C. (2002). Effect of Moisture on the Surface Free Energy and Acid–Base Properties of Mineral Oxides. *J. Chromatogr. A*, Vol.969, No.1–2, pp. 59–72.

Wuhan University (Yang, D. Meng F. Pan Z. Ni Q. & Li G. Ed). (2000). *Analytical Chemistry*, 4th edition, p. 339 and pp. 184–5, Higher Education Press, Beijing, (Chinese).

Vidal, K. Lezama, L. Arriortua, M. Rojo, T. Gutie'rrez, J. & Barandiarán, J. (2005). Magnetic Characterization of $Nd0.8Sr0.2(Mn_{1-x}Co_x)O_3$ Perovskites," *J. Magn. Magn. Mater.*, Vol.290–291, No.4, pp. 914–6.

Yang, F. Zhang, X. Mao, X & Gan, F. (2007). Synthesis and Characterization of Highly Dispersed Antimony-Doped Stannic Hydroxide Nanoparticles: Effects of the Azeotropic Solvents to Remove Water on the Properties and Microstructures of the Nanoparticles, *J. Am. Ceram. Soc.*, Vol.90, No.4, pp. 1019–1028.

Yoffe, D. (1993). Low-Dimensional Systems: Quantum Size Effects and Electronic Properties of Semiconductor Microcrystallites (Zero-Dimensional Systems) and Some Quasi-Two-Dimensional Systems. *Adv. Phys.*, Vol.42, No.2, pp. 173–266.

Zhang, X. (2005). The Preparation and Electrochemical Properties of Sb-Doped Stannic Oxide Nanocrystal. Doctoral Dissertation, Wuhan University, Wuhan, Hubei, China..

Zhang, X. Fu, L. Zhang, M. & Guo, Z. (2008). Solution Method for Synthesizing Nanocrystal of Metal Oxide, *J. Material & Engineering* (China), No.10, pp. 157-163.

Zhang, X. Liang, H & Gan, F. (2006). Novel Anion Exchange Method for Exact Antimony Doping Control of Stannic Oxide Nanocrystal Powder. J. Am. Ceram. Soc., Vol.89, No.3. pp. 792-798

Zhang, X. Guo, Z. Zhang, M. & Fu, L. (2009). Parameters of Condensation of Sb-Doped Stannic Hydroxide and Their Correlation with Nano-Crystalline Sb-Doped Stannic Oxide *J. Sci. Conf. Proc.* Vol.1, pp. 258–267

Zhang, X. Zhou, R. Liu, P. Fu, L. Lan, X. & Gong, G. (2011). Improvement of the Antibacterial Activity of Nanocrystalline Zinc Oxide by Doping Mg (II) or Sb (III) Int. *J. Appl. Ceram. Technol.*, Vol.8, No.5, pp. 1087–1098